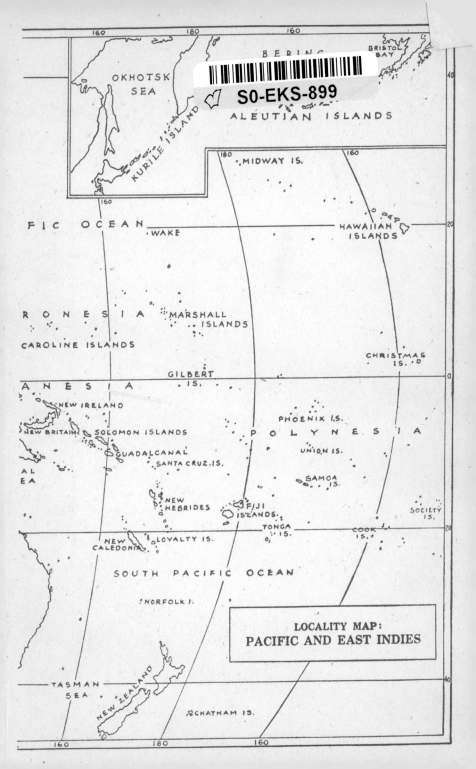

LOCALITY MAP: PACIFIC AND EAST INDIES

PLANT LIFE OF THE PACIFIC WORLD

THE PACIFIC WORLD SERIES

Under the Auspices of The American Committee
for International Wild Life Protection

PUBLICATION COMMITTEE:

Fairfield Osborn, *Chairman* Robert Cushman Murphy
Harold E. Anthony   Edward M. Weyer, Jr.
William Beebe   Childs Frick (*ex officio*)

MAMMALS of the PACIFIC WORLD
 by T. D. CARTER, J. E. HILL and G. H. H. TATE

INSECTS of the PACIFIC WORLD
 by C. H. CURRAN

NATIVE PEOPLES of the PACIFIC WORLD
 by FELIX M. KEESING

REPTILES of the PACIFIC WORLD
 by ARTHUR LOVERIDGE

PLANT LIFE of the PACIFIC WORLD
 by E. D. MERRILL

FISHES and SHELLS of the PACIFIC WORLD
 by JOHN T. NICHOLS and PAUL BARTSCH

THE PACIFIC WORLD
edited by FAIRFIELD OSBORN (W. W. Norton and Co., Inc.)

PLANT LIFE OF THE PACIFIC WORLD

ELMER D. MERRILL
ADMINISTRATOR OF BOTANICAL COLLECTIONS
DIRECTOR OF THE ARNOLD ARBORETUM, AND
ARNOLD PROFESSOR OF BOTANY
HARVARD UNIVERSITY

———————— 1945 ————————
THE MACMILLAN COMPANY – NEW YORK

Copyright, 1945, by
E. D. MERRILL.

All rights reserved—no part of this book may be reproduced in any form without permission in writing from the publisher, except by a reviewer who wishes to quote brief passages in connection with a review written for inclusion in magazine or newspaper.

PRINTED IN THE UNITED STATES OF AMERICA

First printing.

Foreword

This unusual book is one of a series describing the natural history and peoples of the Pacific Ocean and of its innumerable islands, large and small. The basin of this great ocean extends approximately halfway around the earth. Some of the islands, including most of the larger ones, that lie in this vast body of water, owe their origins to the nearby continents; such islands were in earlier times actually a part of their neighboring continents. In other cases the islands arose by powerful geologic upthrusts, including volcanic eruptions, from the very bed of the ocean itself. Because of these different origins, the living things found upon the islands are of infinite variety and interest. This holds true not only of the human beings and other mammals, of the birds, reptiles, and insects dealt with in other books of this series, but also of the plant life described in this volume.

The Pacific World Series has been sponsored, nurtured, and created by men representing nine great American educational and scientific institutions. The original impulse or idea for the preparation of the series came from the American Committee for International Wild Life Protection. This committee not only hoped but believed that a fuller understanding of the wonder and beauty of natural things by those in the Armed Services in the present great war, as well as by their families and friends at home, would create a desire to protect and conserve the natural life of the islands, which, if once destroyed, can never be replaced.

In especial regard to this book on plant life, it is hoped that the information it contains may stimulate visitors to note the natural riches around them, and to collect and prepare specimens for shipment to scientific institutions at home for identification, so that gradually our knowledge of the plant resources of the Pacific may be widened. For although 45,000 different kinds of plants are now known in this area, the author of this book, and others wise in the lore of Pacific plant life, believe that many thousands of kinds still await discovery and study.

In the first line of this Foreword the adjective *unusual* was used in connection with this book. At best it is but a weak term to describe an extraordinary accomplishment. The author attempted almost the impossible when he undertook to present, in relatively limited space, so intricate a subject. He has been extraordinarily successful for two reasons: first, because of his own remarkable powers of observation; second, because for forty-two years he has studied the plants of the Pacific and the bordering continental lands to the west, and consequently he writes from the richness of personal experience. For nearly twenty-two years he resided in the Philippines, and while there explored all parts of that archipelago, as well as parts of Java, Borneo, the Malay Peninsula, and southern and eastern China. He has been the discoverer of many new genera of plants and has actually named and described approximately 4,000 species of plants from these regions, which had not previously been known to science. This book is not an encyclopedia. The author has somehow avoided a dry, systematic presentation. Instead, as a review of his chapter headings will indicate, he has created an embracing and over-all view of the subject, harmonious in structure, expressed in broad overtones, yet supported by adequate detail to bring to the reader a clear understanding of the plant life of these far-flung islands. Of him no one can say that he "cannot see the forest for the trees."

FAIRFIELD OSBORN, *President,*
New York Zoological Society.

Contents

	Foreword	v
	Introduction	xi
1.	The Safe Forests and Jungles of the Tropics	1
2.	General Principles of Botanical Classification	9
3.	Plants of the Seashore	27
4.	The Mangrove Forest	49
5.	The Secondary Forests and Open Grass-lands	59
6.	The Primary Forest	71
7.	Noteworthy Plants of Special Interest	93
8.	Weeds and Their Significance	118
9.	The Cultivated Plants	147
10.	Jungle Foods	178
11.	Problems of Malaysian Plant Distribution	193
12.	Problems of Polynesian Plant Distribution	205
13.	The Significance of Certain Local Plant Names	220
14.	Notes on Specific Islands and Island Groups	232
15.	Notes on Botanical History, Exploration, and Bibliography	242
16.	Simple Directions for Preparing Botanical Specimens	257
	Botanical Arrangement of the Species	263
	Glossary	277
	Index	283

Illustrations

FIGURE		PAGE
1	Pitcher plants	3
2–4	Poisonous plants	6
5	Schematic arrangement of the plant kingdom	12
6	Leaf shapes	20
7	Leaf tips and bases	21
8	Leaf margins and forms	22
9	Floral parts and floral types	23
10	Types of inflorescences	24
11–51	Strand plants	40–48
52–54	Typical mangrove trees	51
55–66	Secondary mangrove forest types	57–58
67–84	Secondary forest types	68–70
85–87	Cauliflory	95
88–96	Symbiosis with ants	97–99
97	Stenophyllous plants	100
98–100	Unusual plants	104
101–111	Strange ferns	107–109
112–189	Weeds	134–146
190–201	Edible plants	167–168
202–207	Edible fruits, seeds and tubers	169
208–212	Edible plants	170
213–218	Edible fruits	171
219–253	Ornamental plants	172–177
254	Chart of the eastern Malaysian region	199
255	Chart showing the attenuation of a large typically Malaysian genus in Polynesia	217
256	A simple botanical press	258

The photograph used in making the jacket for the book is used through the courtesy of the U. S. Navy, File No. 8710.

The end paper is taken from *Mammals of the Pacific World* by Messrs. Carter, Hill, and Tate.

Introduction

Collectors and connoisseurs, whether interested in plants, insects, shells, birds, postage stamps, rare books, antiques, objects of art, or any other category, realize from their own experience the thrill that is felt when a long-sought-for item is first located. One of the most striking published descriptions of this emotion is the anecdote related by Conrad in his *Lord Jim* regarding Mr. Stein, a merchant in Java, one of the characters in the book. As narrated, the anecdote opens with the statement: "You don't know what it is for a collector to capture such a rare specimen. You can't know." When a young man, Stein had served as an assistant to a famous entomologist who for years had sought in vain for a particularly rare, large, and brilliantly colored butterfly. Later when Stein had settled in Celebes, he received a message from a local chief, requesting him to come to a certain place for a conference. En route he was ambushed by a band of armed natives, and after the first volley of rifle shots seven men rushed at him to complete the kill. Waiting until they were very close, Stein shot three of them, and the rest escaped.

This alone should have been venture enough, but as he was examining the face of one of the slain men for any possible signs of life, he noted a faint shadow, and glancing up, he saw the marvelous butterfly for which he and his former employer had so long hunted in vain. It gradually descended and finally alighted some distance ahead on the trail. Dismounting from his horse, he approached it very carefully, hat in hand. On capturing it, he expressed his sensations thus: "I shook like a leaf with excitement, and when I opened those beautiful wings and made sure what a rare and so extraordinary perfect specimen I had, my head went round and my legs became so weak with emotion that I had to sit on the ground. I had greatly desired to possess myself of a specimen of that species when collecting for the professor. I took long journeys and underwent privations; I had dreamed of him in my sleep, and here suddenly I had him in my fingers—for myself."

Such emotions may be the experience of some who peruse this book when, unexpectedly and perhaps under strange circumstances, they encounter some of the more rare or striking plants herein discussed, or other forms of life mentioned in companion volumes.

On first observing the tropical scene, the individual is impressed by several facts, and may be overwhelmed by the richness and complexity of the plant life. The very exuberance of the tropical vegetation and the overpowering mass of green, in general unrelieved by quantities of showy flowers and with palms, bamboos, large trees, and various other types of tropical plants dominant in places, may be discouraging to the novice. Meadows and fields covered with masses of brilliant flowers do not exist, for the bulk of the vegetation in open places consists of coarse grasses. Even though thoroughly familiar with the flora of the temperate region whence he came, the observer may see little or nothing with which he was acquainted at home, other than certain cultivated food plants, an occasional weed, or perhaps some tropical species that has become more or less common in cultivation in northern regions. If he does detect the lowly purslane, which is an ubiquitous weed, or in wet places a cat-tail similar to those species that dominate our own swampy areas, he may greet them as long-lost friends. While in places along the forested margins of rivers or along the seashore he may observe lianas or even scattered trees with numerous large and brilliantly colored flowers, such plants will in general be few. In most of the tropical Pacific region, the species with masses of small colored flowers or with large showy blooms that will be noted in cultivation in the settled areas, originated in other parts of the world, whether they be vines, shrubs, or trees. In modern times such regions as Madagascar, tropical Africa, and tropical America have contributed their quotas of showy exotics to the ornamental plants of Malaysia and to the smaller islands of the Pacific. These exotics naturally will not be found in the jungles and in the forests, but only in settled localities.

An attempt is made in this book to cover a vast region, extending from the Aleutians in the north to Hawaii, the Marquesas Islands, and the Galápagos in the east, westward across the Pacific to include all of the Malay Archipelago and the Papuan region as far to the south as New Caledonia, as well as, with brief and partly bibliographic notes, the islands adjacent to eastern Asia, from the Kuriles to Formosa and Hainan. In the text the term Malaya is, in general, confined to the Malay Peninsula, but the term Malaysia covers the entire Malay Archipelago, including the Philippines, that great group of islands—the largest archipelago in the world—situated between southern Asia and northern Australia. The term Sunda Islands covers only western Malaysia, Sumatra, Java, and Borneo and the adjacent smaller islands. Papuasia indicates New Guinea and its adjacent islands. Micronesia appertains to the very numerous small islands extending from the Caroline and the Marianas Islands to the Marshall Islands, essentially the area covered by the Japanese Mandated territory; Polynesia applies to the islands of the central and eastern Pacific, and if, on occasion, the term Melanesia is used, this indicates those islands in the southwestern Pacific region extending from the islands adjacent to New Guinea to Fiji.

In the vast region covered by this work there are certainly in excess of 50,000 different species of higher plants. True, many regions are characterized by strictly limited and relatively poor types of plant life, such as the small low islands in the Pacific Ocean itself, and those in the extreme north; yet others, such as the larger islands in Melanesia, Papuasia, and Malaysia, support an extraordinarily rich plant life. While many parts of the area have been extensively and intensively explored from a botanical standpoint, few have been covered by exhaustive descriptive manuals or even by published lists of the known species (see p. 248).

Some idea of the great plant wealth of the Indies may be gained from the statement that in excess of 5,000 different species of orchids and 2,000 different species of ferns and fern allies are already known from these widely scattered islands. In the Philippines alone, some 9,500 species of higher plants are now known, and yet the land area of the Philippines merely approximates that of the New England States and New York combined. These parts of the northeastern United States support only about one-third as many plant species as the Philippines. In a single volume published about thirty years ago, no less than 1,153 new species of orchids were described from northeastern New Guinea (German New Guinea or Kaiser Wilhelm Land) alone; yet in all of North America north of Mexico, only about 150 orchid species occur. In many parts of such islands and island groups as Sumatra, Borneo, Celebes, the Moluccas, New Guinea, the Solomon Islands, Bismarck Archipelago, and the New Hebrides, a distinctly high percentage of all the species occurring are as yet unnamed and undescribed, for vast areas are still, from a botanical standpoint, literally *terræ incognitæ*.

Naturally, a small volume on the plant life of the islands of the Pacific could not be a complete treatise even for any part of this vast area. A descriptive manual alone would be encyclopedic in nature, and such a work would defeat the objectives of this series of publications. It could not be even a partial list of the known species, for such a list would be of little practical value even to a trained and experienced botanist.

For the estimated 50,000 different species of plants that occur in the regions covered—and this number is being constantly increased as exploration progresses—more than 2,500 different genera are represented. For the known species, in excess of 45,000, local or native plant names are already recorded in association with their corresponding Latin binomials, and many thousands of local names still remain to be recorded. This total is not surprising when one realizes how rich the floras are, and further, that within the area involved perhaps as many as 500 different languages or dialects are spoken. It scarcely needs to be pointed out that most of these numerous vernacular names are in very local use, and hence they are merely of local value to the man in the field. Few of the numerous species have any common English names, the result being that there was no choice but to use the technical Latin names for the selected species mentioned in this book.

In considering what to include and what to exclude, the first decision made was to confine the species discussed to those of extraordinarily wide geographic distribution, and again, those that are the most common ones. The second decision was to utilize the available space for generalized treatments of certain definite types of vegetation in the mass and various correlated subjects as detailed in the text.

The work has not been prepared for the professional botanist. Rather, an attempt has been made to keep in mind the needs of the lay reader. From the very nature of the case, some of the chapters have to be factual, and such chapters will, in general, be of greatest interest to those concerned about plants and plant names. If this volume merely suggests fields of potential interest to individuals far from home and often stationed in strange surroundings, it will have served its purpose. After all, we should not forget that plant life is the basis of all other life, for as the Good Book says, "All flesh is grass," which is merely a statement of the obvious fact that all animal life, including man, is dependent directly or indirectly on plants. Thus there is in all regions an intimate correlation between the plant life and the animal life of each area; and generally speaking, where vegetation is highly developed there is a corresponding richness in animal life.

Attention is called to the recently published *The Pacific World*,* the forerunner of this series of texts, edited by Fairfield Osborn, President of the New York Zoological Society, an excellent introduction to a study of the Pacific region.

Two other recent publications are perhaps worthy of note, my own "Emergency Food Plants and Poisonous Plants of the Islands of the Pacific" † and the Kraemer "Native Woods for Construction Purposes in the Western Pacific Region." ‡

The illustrations, with one exception, were prepared by Miss Martha Suttis. One map was drawn by Mr. Gordon Dillon.

* OSBORN, FAIRFIELD (Editor). *The Pacific World*. Its vast distances, its lands and the life upon them, and its peoples, pp. 1–218, illus., 1944. The W. W. Norton and Company, New York. Fighting Forces edition (unabridged). The Infantry Journal, Washington, D. C., 1944. Copyrighted by the American Committee for International Wild Life Protection.

† War Department, Technical Manual 10-420, pp. 1–149, figs. 1–113, 1943. Government Printing Office, Washington, D. C. Superintendent of Documents, price 15 cents.

‡ Bureau of Yards and Docks, Department of the Navy, Washington, D. C., pp. 1–382, figs. 11–101, 3 maps, 1944 (Restricted Publication).

1

The Safe Forests and Jungles of the Tropics

Strange tales are told regarding the perils of the tropical forests, such as the utterly fabulous ones about the deadly upas-tree (*Antiaris toxicaria*) of Malaysia and the purely imaginary "man-eating tree" of Madagascar. And if the Sunday supplements are to be believed, even the Philippine forests harbor ferocious man- (or maiden-) eating trees. But, perhaps unfortunately for science, not one of the hundreds of plant collectors, experienced explorers, botanists, and highly trained foresters and forest rangers who have traversed these forests from end to end, year after year for the last four decades, not to mention their less numerous predecessors, ever encountered or even heard of these vegetable monstrosities. Tales regarding man-eating trees are pure fiction.

THE UPAS-TREE

It is true that the upas-tree does exist in Malaysia, and that its milky sap is very poisonous once it is introduced into the bloodstream. Even today the aborigines use the milky juice to poison their darts and arrows, but the plant itself may be handled with absolute safety. Among the fantastic tales about this tree are those to the effect that it was so dangerous that whenever its sap was needed to poison darts or arrows, only condemned criminals were told off to gather it; that birds perching on its branches, and animals passing underneath the trees, immediately died, and that the ground under the trees was covered with the bones of these innocent victims. A wholly imaginary and grossly exaggerated account of the upas-tree was published in London in 1783, and the lurid tales so impressed Erasmus Darwin that, with poetic license, he incorporated the more striking of these passages in his poem *Loves of the Plants,* a work that became widely popular; and thus the fabled upas-tree became "damn'd to everlasting fame."

HOW DID PRIMITIVE MAN DISCOVER THE POISONOUS PROPERTIES OF CERTAIN PLANTS?

Manifestly, trial and error played a most important part in determining what could and could not be eaten with safety. If an individual ate certain fruits or seeds and then died, it soon became generally known that such fruits or seeds should be avoided, and this information became traditional. How did primitive man discover, for example, the poisonous properties of the milky sap of the upas-tree? The explanation here is relatively simple. Throughout Malaysia occurs a woody vine (*Parameria lævigata*) of the Apocynaceæ, with the latex containing a considerable amount of rubber; this latex is innocuous. It is extensively used when a cut or abrasion is to be treated, merely by permitting the milky sap to drip from cut ends of the vine upon the injured surface; because of the presence of rubber in this milky sap, a film or a sort of "new skin" is formed over the injured part. Secondary infection is thus largely eliminated and the wound soon heals. In the Philippines this plant is widely known as *dúgtong áhas,* which on analysis impresses one as an utterly meaningless name, for the first word means "to cut," and the second, "snake." The milky sap is considered to be so efficacious in treating wounds that many natives firmly believe that if a snake be cut in two parts, the cut parts joined, and the wound smeared with the latex of *dúgtong áhas,* the snake will revive and go on its way rejoicing in its new life—hence this rather absurd but widely used plant name. In the dim past, some native with an experimental type of mind reasoned that if the latex of *dúgtong áhas* could be used, then any milky sap might be substituted. But he, unfortunately, tried the abundant milky sap of the upas-tree. With its powerful poison thus entering his bloodstream, the primitive experimenter promptly died. The very poisonous property of this particular milky sap once established, it was only a simple step to smear the points of arrows, darts, and spears with the upas latex for the purpose of killing birds, animals, and enemies. Just a little simple experimentation resulting in the discovery of a poison potent in the bloodstream.

THE FICTITIOUS MAN-EATING TREE

Even more strange are the tales regarding the man-eating tree, but all of these are utterly fictitious, without even vestiges of a basis of fact. Actually there is nothing closer to a "man-eating tree" anywhere in the world than the utterly harmless and very small sundew plants (*Drosera*) and the curious allied North American Venus fly-trap (*Dionæa muscipula*). The few representatives of *Drosera* that occur in Malaysia are very similar in size and appearance to the native species of Europe and the United States. Their leaves can and do capture and digest mosquitoes and small flies, but nothing larger.

INSECTIVOROUS PLANTS

The most spectacular insectivorous plants in Malaysia are the rather numerous representatives of the pitcher-plant (*Ne-*

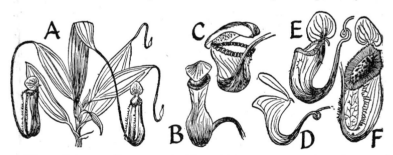

Fig. 1. PITCHER PLANTS: A. *Nepenthes papuana*; B. *N. clipeata*; C. *N. ephippiata*; D. *N. inermis*; E. *N. treubiana*; F. *N. merrilliana*.

penthes, fig. 1). Approximately eighty-five different species are known within the limits of its generic range, which extends from Madagascar to India and southeastern China, southward and eastward through Malaysia to New Caledonia. Most of the species are Malaysian; some occur at sea-level, others on the higher mountains. They are chiefly vines, and the variously colored and often mottled pitchers are in general reddish to purplish. The species having the largest pitchers, which may hold a quart or more of water, are indigenous in Borneo and the Philippines. The pitchers are so balanced on their twining stalks that when too much water accumulates in them, they tilt and spill a part of it. They are provided with somewhat protective lids, and on the inside, near the top, is a series of reflexed or downward-projecting small spines or teeth which prevent insects from escaping. Eventually, insects that enter the pitchers drown and decay, the nutrients thus provided being absorbed by the plants. It is a very far cry indeed from the utterly harmless sundews and pitcher-plants to the fabulous man-eating tree!

CONTACT POISONS

It is true that there are certain shrubs and trees in Malaysia that are definite contact poisons, quite like our common poison oak and poison ivy (*Rhus*). Most of these contact poison plants belong in the same natural family, the Anacardiaceæ, as does *Rhus*; representative poisonous genera in the Old World tropics are *Semecarpus,* fig. 4, *Gluta, Swintonia, Melanorrhœa, Melanochyla,* and even some species of mango (*Mangifera*). These, with the exception of a few shrubby species of *Semecarpus,* are all trees. Various representatives of these dissimilar genera are widely known in Malaysia as *réngas,* or in the Philippine Islands under the cognate form *ligas.* Here the native name does not indicate any similarity between what are actually very different kinds of plants, but rather a group of plants with one character in common—in this case, contact poisons. The causative agent producing the characteristic skin eruptions and intense itching is a non-volatile irritating oil comparable to that which is present in poison oak and poison ivy; hence the treatment should be the same as for that of *Rhus* poisoning at home. All of these poisonous plants have a sap that soon turns black on exposure to the air; it is the sap, sometimes slightly milky, that carries the poison.

The chances of coming in contact with these poisonous plants in the tropical forests are distinctly less than with poison ivy at home. Generally speaking, it will be only those who fell trees and who inadvertently smear the sap on their skin who will be affected. Yet a number of serious contact poison cases have been reported from the Southwest Pacific region. In one case the individual was very badly poisoned merely by coming in contact with freshly sawed lumber that was used for the construction of a shelter. In the region concerned, this was probably a species of *Semecarpus,* for representatives of the other poisonous genera, other than *Mangifera,* are not known to occur in that part of the world; and there is no available evidence that the few native species of the last genus in New Guinea and in the Solomon Islands are actually poisonous.

It is a rather curious fact that the mature fruits or seeds of many of these Malaysian contact poison trees may be eaten with entire safety. Fortunately again, there seems to be a fairly high percentage of immunity to the poison; some individuals are highly susceptible to it, others not. However, the abundant milky sap of a very common seaside shrub or small tree (*Excœcaria agallocha,* fig. 31), of the euphorbia family, is reported to cause very severe conjunctivitis in contact with the mucous membranes about the eye.

STINGING PLANTS

The tree nettles, of the genus *Laportea,* fig. 3, all shrubs or small trees, are sometimes common in the forests and thickets. Some of them have rather large and alluringly soft leaves. The genus is very closely allied to *Urtica,* to which our relatively harmless stinging nettles belong. When one touches the leaves of these tree nettles even gently, the immediate sensation is almost like that of having touched very hot iron. The sap in the slender stinging hairs is intensely irritating, due to the presence of formic acid, and it causes extensive water blisters and much pain. However, the stinging of these supernettles is normally not dangerous, and not even a cub Boy Scout would knowingly touch a plant a second time. Other plants, chiefly vines, bear stinging hairs, but these are for the most part not poisonous. The commonest source of these needle-like, stiff stinging hairs is the often peculiarly sculptured pods of certain

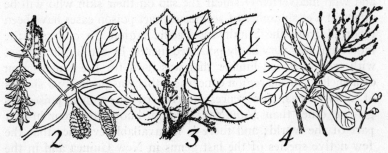

POISONOUS PLANTS: Fig. 2. *Mucuna*; Fig. 3. *Laportea*; Fig. 4. *Semecarpus cuneifolius*.

species of the genus *Mucuna,* fig. 2, of the bean family, but these hairs are merely mechanical in action—irritating but not dangerous.

TRAVELERS' TALES

While the relatively few, grossly exaggerated tales regarding the dangerous plants of the tropics are still repeated, such as those about the completely fictitious man-eating tree and the existent but not so deadly upas-tree, it is in the animal kingdom that we note the worst overstatements. Large predaceous animals in general do not exist within the Malaysian region, except for the tiger in some of the Sunda Islands and in the Malay Peninsula, although at times some of the large herbivorous mammals may be a source of danger. This is the case with the domesticated carabao or water buffalo, and particularly the feral type; but this animal is no more to be feared than is the domesticated bull with us. The larger mammals of the region such as the orangoutang, wild hog, deer, and others, are timid and normally are not to be thought of as dangerous, although large ruminants, like the sladang and the timarau, are to be avoided in such places as these animals occur.

Most thoroughly to be discounted are the wildly exaggerated stories, repeated over and over again, regarding "the snake-infested jungles of the tropics." True, snakes do occur, and some of them are poisonous; others, such as the python, are very large in size. But to find snakes, one must know when and where to look for them, for, like other denizens of the jungle, they are timid and slip quietly away at the slightest disturbance. Even the largest pythons are not dangerous to man. The ordinary sojourner in the jungles of the Old World tropics will indeed be fortunate if he averages seeing a snake a week, and one may pass week after week in the forests and jungles and never encounter a single serpent. Most travelers' tales regarding snakes in the tropics are not to be trusted, and these stories, told and retold for effect, lose nothing in the telling.

Audience hypnotizers use the half-truth method, as, for example, the tale regarding the nesting habits of the hornbill. It is true that the female bird, during the incubation period, remains on the nest in a hollow tree with the opening closed by mud in such a manner that she cannot escape, for the male bird leaves an opening just large enough for her bill to protrude. During the incubation period the male bird feeds the female with various fruits, doubtless including those of the strychnine plant. The entertaining lecturer is careful not to state that the bird eats only the pulp and not the seeds, for even man can with entire safety and considerable enjoyment eat *Strychnos* fruit pulp since it is not only innocuous but is well flavored. The hornbill no more eats the fairly large, hard seeds than does the ordinary man eat the pits of peaches, prunes, and cherries; but the half truth that the male hornbill feeds the female on *Strychnos* fruits always succeeds in making an audience gasp, for everybody knows that strychnine is a deadly poison.

The less said of the famous or infamous tree-climbing fish of Malaya the better, for this insignificant little gobie is not the large and spectacular fish that is inferred. It is never found in the forests, but does occur by thousands and tens of thousands on rocks and on the roots and trunks of mangrove trees along the seashore in quiet places, and sometimes it may climb as much as a foot above the surface of the water. It is so insignificant that the average person would never glance at it a second time; and yet it too serves as audience bait.

DEFLATION

Would that it were possible to prick the myriads of bubbles based on half-truths, inferences, and deliberate falsehoods on which our public has been regaled regarding the non-existent terrors of the tropical jungles! These nature fakers with their ill-conceived and misdirected fables about the dangers they met (and always overcame), who write imaginary stories and who hypnotize audiences with overdrawn statements regarding their ventures and misadventures, interspersed with half-baked observations on life in the imaginary snake-infested jungles of the tropics, have rendered a distinct disservice to our public.

After all, with reasonable precautions, and under anything remotely approaching normal conditions, the jungles and forests of the Old World tropics are safe places in which to live and work. There will be discomfort in stormy or rainy weather; there will be high temperature, but nothing as bad as conditions that prevail in the summer in some of our Atlantic seaboard and inland cities; there will be high humidity; there will be many mosquitoes, and through these the ever-present possibility of contracting malaria; there will be plenty of ants and various other insects; there will be land leeches within the forests in wet weather; and there may be some scorpions and centipedes. But on the whole, forests in the tropics are little more terrifying than are those at home. Bear in mind that many millions of native peoples live and have their being in these supposedly ultra-dangerous and much maligned regions, that for centuries Caucasians have visited the islands, and that hundreds of thousands of these sojourners have lived there for extended periods, many for most or even all of their lives. Millions of people living in the tropics, the successors of untold millions who preceded them, prove, if proof be needed, that the tropics are reasonably safe.

2

General Principles of Botanical Classification

Under normal conditions the individual interested in the names, uses, relationships, and classification of plants would refer to standard texts, and were he in residence near a well-equipped botanical institution he would seek advice and assistance there. In the Old World tropics there are very few botanical centers where ample library and reference collections are available; they are practically confined to Honolulu, Manila, Singapore, Buitenzorg, and Calcutta.

For the person interested in plants, there is always the problem of discovering what has been published regarding the vegetation of any particular region, whether or not such publications are available, and where copies may be secured. For this reason alone it was decided to include certain bibliographic data in this work, merely to indicate in a most general way what is actually available in printed form. (See Chapter 15.)

To make this volume more complete in itself, and particularly to meet the needs and requirements of individuals not trained in botany, this simple chapter was prepared. There are relatively few botanists in any country who are at all familiar with the exceedingly rich vegetation of the Oriental tropics, but their services are available to any and all who may seek advice and assistance. For sources of such assistance see Chapter 16.

MAJOR DIVISIONS OF THE PLANT KINGDOM

In broad outlines the plant kingdom is divided into four major groups. The most primitive is the *Thallophyta,* or cellular cryptogams, including those plants which are reproduced by simple division or through the production of spores. This primitive group includes the fission plants (*Schizophyta*), such as the bacteria and the blue-green algæ; the true *Algæ,* including the green (*Chlorophyceæ*), red (*Rhodophyceæ*), and brown (*Phæophyceæ*); and the *Fungi,* the latter consisting wholly of *saprophytes* (living on decaying matter), or *parasites* (living on other plants or animals).

The second major group is the bryophytes (*Bryophyta*), composed of the true mosses (*Musci*), and the liverworts (*Hepaticæ* and *Jungermanniaceæ*). In these two groups there is no development of woody tissues or vascular elements; hence they are both included among the cellular cryptogams. They are not further considered in this treatise, but individuals interested in them may secure advice and assistance through correspondence.

THE FERNS

The third major group includes the ferns and fern allies, often referred to as the vascular cryptogams, all much more highly organized plants than are the cellular cryptogams because in their tissues the vascular elements are more or less developed. Like the cellular cryptogams, they are reproduced by spores. Technically the group is known as the *Pteridophyta,* which merely means fern plants. At least 2,000 different species occur in the region covered by this volume, but only a few of the more common or more striking ones are mentioned in the text. Here again, advice and assistance will be made available to all who may be interested in these most attractive plants.

THE FLOWERING OR SEED-PRODUCING PLANTS

The fourth and most highly developed group, the flowering plants (*Phanerogamia* or *Spermatophyta*), includes those which produce true flowers and which are reproduced by seeds. These plants form the bulk of our present-day vegetation. There are two great subdivisions, the *Gymnospermæ* and the *Angiospermæ.* The first is the more primitive, and in earlier geologic times dominated the vegetation of the earth. The technical name means "naked seeds," for not only are the seeds not enclosed in seed-vessels (fruits), but the ovules themselves are also naked, and the stigma is absent. This group includes the cycads (*Cycadaceæ*) and the *Coniferæ*, the latter assemblage including our characteristic northern evergreens, such as the pine, spruce, fir, hemlock, cedar, juniper, and others. There are relatively few representatives of this group in the tropics.

THE MONOCOTYLEDONS AND THE DICOTYLEDONS

The second group of the true flowering and seed-producing plants is known as the *Angiospermæ* (seeds produced in closed seed-vessels or fruits). These plants are characterized by the ovules being developed within closed ovaries, the ovary always provided with a characteristic stigma, and the seeds developed within fruits of one type or another. This great group dominates the tropical vegetation and vast areas in the cooler parts of the world. It is divided into two great subdivisions, the *Monocotyledoneæ* and the *Dicotyledoneæ*. The first group, frequently spoken of as the monocotyledons, is strikingly characterized by its seeds having a single *cotyledon* (seed leaf), its leaves for the most part having parallel veins, and the flower parts normally in threes or multiples of three; it is represented by such natural groups (families) as the palms (*Palmæ*), lilies (*Liliaceæ*), grasses (*Gramineæ*), sedges (*Cyperaceæ*), and orchids (*Orchidaceæ*).

The second group, the *Dicotyledoneæ* or dicotyledons, is characterized by its seeds having two *cotyledons* (seed leaves), as in the common bean, its floral parts mostly in fives, and its leaves with a characteristic net venation. This is the group that contains most of the trees, shrubs, and vines, other than the palms in the monocotyledons, and the common evergreen trees mentioned above that are so dominant in many cooler regions, and also many herbs.

While this general scheme of classification of major groups is not the latest one proposed, it is convenient in bringing the lesser categories into a few easily defined and natural assemblages, and is the one now most widely used. There is no such thing as a possible linear arrangement whereby a primitive group

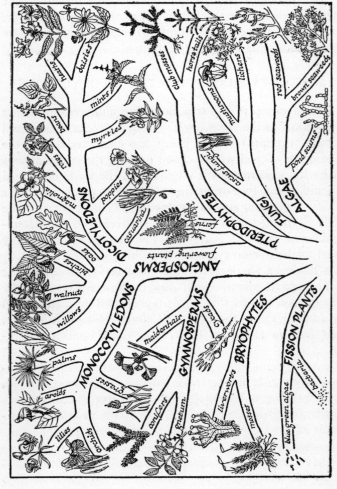

Fig. 5. Schematic arrangement of the plant kingdom, indicating the major groups and selected families.

is followed by the next higher one, and so in sequence from the very lowest to the very highest plants in the plant kingdom. Rather, in evolution certain developments have attained dead ends, and from time to time in the history of the plant world new trends have been initiated. The scheme can best be demonstrated by comparing the plant kingdom to a much-branched tree, with the primitive form or forms at the base, the lower branches the more highly developed groups, the highest main branches the most highly developed plants, and finally the ultimate branchlets of these higher branches representatives of the natural families into which the entire group has been subdivided (fig. 5). In spite of the great diversity in plants, there is a remarkable series of steps from the lowest to the highest, with here and there apparently retardation or even degeneration, which is exactly what would be expected if the generally accepted theory of evolution be true.

PLANT FAMILIES

The major divisions and subdivisions of the plant kingdom as briefly outlined above are again subdivided into families. Botanists have attempted to arrange these families in accordance with their relationships and inter-relationships. Thus in any standard descriptive manual or *flora* (by the word *flora* is meant all the plants of a given region, or a descriptive work wherein they are described), the sequence of arrangement will be that of some standard work in which all the groups of plants for the entire world are treated. Several different systems of arrangement of families have been proposed, but it is hardly necessary to discuss these technicalities here.

A family is an assemblage of minor groups (genera), all of which have certain characters in common, but a combination of characters that is not found in any other group. Many of these families are very easily recognized at sight, even by a layman with no botanical knowledge, such as the monocotyledonous ones above mentioned, and such strongly marked groups in the dicotyledons as the cacti (*Cactaceæ*), the rose family (*Rosaceæ*), the bean family (*Leguminosæ*), the squash family (*Cucurbitaceæ*), the mint family (*Labiatæ*), the sunflower family (*Compositæ*), and many others. For the entire world approximately 300 different families are currently recognized, with about 10,000 genera for the higher groups of plants alone. A rough estimate is that in these higher groups in excess of 200,000 different species are known, and the total increases as exploration progresses.

Even an individual untrained in botany will at sight recognize certain relationships in plants, perhaps not realizing that what he observes in common as between otherwise often rather dissimilar plants may really represent the technical characters utilized by botanists in delimiting and characterizing natural families. The most casual observer will recognize in such widely cultivated species as the squash, pumpkin, melon, watermelon, gourd, cucumber, and various wild plants, certain characters in common, such as the general appearance of the flowers, the cut of the leaf, the presence of tendrils, the characteristic fruits and seeds, and the habit, in that all of these are vines. There will be differences in size, color, shape, and other characters of the various organs, yet through all of these diverse plants will run the same general pattern. Here we are dealing with a very natural family, for the peculiar combination of characters that delimit the *Cucurbitaceæ* will not be found elsewhere in the plant kingdom.

Another striking example is the bean family (*Leguminosæ*). No matter what the differences may be as between its very numerous genera—for the differences are very great in habit, leaves, flowers, and fruits—there is one constant character that persists and does not occur elsewhere. In all the numerous genera and species, the seeds (beans) are produced within a characteristic one-celled fruit, a true pod, that may split open along one or both sides, or may remain closed. The numerous representatives vary from insignificant to large herbs, slender vines, great lianas, shrubs, and often giant trees, particularly in the tropics; yet this fruit character persists and occurs nowhere else in the plant kingdom.

THE GENUS

Under the family, as a matter of convenience, genera have been established. An attempt has been made to bring together all species that have certain characters in common, those of a secondary nature as compared with the wider characters used to delimit the family. Thus the oaks (*Quercus*), the chestnuts (*Castanea*), and the beeches (*Fagus*), form separate genera within the family *Fagaceæ*. In all there are certain vegetative, floral, and other characters in common, yet one immediately recognizes an oak (*Quercus*) by the distinctive acorn, a chestnut (*Castanea*) by its spiny burr containing several fruits, and a beech (*Fagus*) by its smaller burr and characteristic three-angled fruits.

If one examines the flowers of the cherry, peach, plum, pear, apple, rose, raspberry, and strawberry, one will note that structurally and in general appearance they have much in common, although their fruits are very different. Those with stony, solitary pits, such as the cherry, peach, and plum, belong in the genus *Prunus*, the rose in the genus *Rosa*, the apple in the genus *Malus*, the pear in the genus *Pyrus*, the raspberry in the genus *Rubus*, and the strawberry in the genus *Fragaria*. These, with many other genera, make up the very natural family *Rosaceæ*.

THE SPECIES

While a natural family may be limited to a single genus, or may be composed of several hundred different genera, so a genus may be *monotypic* (composed of a single species), or *polytypic* (composed of many species). The species is a concept based on the observed fact that normally the different kinds of plants are reproduced in nature in kind, all descendants of a certain plant or plants having, within narrow limits, the characters of their parents, and the fact that this process continues generation after generation and century after century. There is, of course, some variability, for, after all, no two plants are exactly alike in absolutely all characters; but in the species concept all descendants of an established or recognized entity are approximately like the parent plant.

THE BINOMIAL

In botany, as in other natural sciences, the binomial (or two-part name) is used to designate individual species. It is composed of a generic and a specific name. One or both of these may be descriptive words, derived from classical sources, Latin or Greek, or Latinized words derived from place names, or the names of individuals, or even in some cases from vernacular names. Take the binomial used to designate the common mangrove tree (*Rhizophora mucronata* Lamarck) as an example. The generic name was derived from two Greek words meaning "root" and "to bear," for all the species of this genus have very strongly developed prop roots. The specific name was derived from the fact that the leaf tip bears a slender short prolongation of the midrib as a distinct mucro. The binomial was originated by Lamarck in 1796.

Thus when we think in terms of plant names, we think of an accepted generic term, plus a specific name, with or without the name of the individual who first published the combination. After all, this is a very simple technique, for it corresponds to such common English names as white oak, red oak, swamp oak, live oak, scrub oak, or red pine, white pine, pitch pine, yellow pine, where the words oak=*Quercus* and pine=*Pinus* represent the generic concept, and the modifying word the species concept. Curiously, this use of binomials in everyday speech literally goes back to the early history of languages, and yet the naturalists did not apply the system in the form of technical names to plants and animals until the middle of the eighteenth century.

ORIGINS OF GENERIC NAMES

Perhaps one could prepare a very interesting essay on the origins and meanings of technical names, but this is scarcely the place to amplify the subject. Many hundreds of generic names, and thousands of specific names, are derived from the names of people. We use in everyday speech such names as *Lobelia, Dahlia, Zinnia, Begonia, Kalmia,* and others, and think nothing of it; yet we are apt to balk at such unfamiliar names as *Bruguiera, Dioscorea, Burmannia, Barringtonia, Lagerstrœmia, Osbeckia, Torenia, Cunninghamia, Clintonia,* and *Bartramia.* Who were Messrs. L'Obel, Dahl, Zinn, Begon, Kalm, Bruguières, Dioscorides, Burman, Barrington, Lagerstrœm, Osbeck, Toren, Cunningham, Clinton, and Bartram? These are the names of certain pioneers who prosecuted important botanical work, sometimes serving only as collectors, or in other cases individuals who deserved well of botanical science because of support provided by them.

A great many generic names are classical, in that the names used today to designate certain groups were the words actually used by the ancient Greeks or Romans as vernacular names for plants known to them, such as *Pinus* for the pine, *Fagus* for the beech, *Quercus* for the oak, *Malus* for the apple, and *Pyrus* for the pear, to mention only a few. Many others are coined names from classical sources, being descriptive of some character of the plant, such as *Epigæa* = "upon" and "earth," from the creeping habit of our common mayflower, or *Helianthus* = "sun" and "flower" for the common sunflower, or *Artocarpus* = "bread" and "fruit" for the tropical tree now so widely known as breadfruit. Still other generic names are either unmodified, or sometimes Latinized vernacular names, such as *Tsuga,* the generic name for the hemlocks, this being the Japanese vernacular name for one of the Asiatic species, or *Palaquium,* derived from the Tagalog *pálak-pálak,* the local name for one of the Philippine gutta-percha trees.

THE OBJECTIVE OF TECHNICAL NAMES

In the biological sciences a fixed name for each separate entity, whether it be genus or species or some other category, is highly essential. A widely distributed tree species such as *Hibiscus tiliaceus,* that occurs naturally along the seashores throughout the tropics, will literally have many scores or perhaps hundreds of native names, for each people and often each tribe will have its own name for this useful tree. Yet among botanists everywhere the simple binomial stands among all civilized peoples as the proper name for this particular tree, no matter whether it be observed in China, in India, in New Guinea, in Africa, or in tropical America. The advantage is a relatively simple name that will be understood in its application among all people, at least the technically trained ones, and in all countries. It so happened that the classical influence of Latin, now a "dead" language, extending downward through the Middle Ages in Europe, became the language of science. As such it was acceptable to all scientists, and has continued to be acceptable to all modern investigators, for it is universally used, whether they be of English, German, Dutch, French, Russian, Chinese, Japanese, or Indian extraction, or the representatives of any other racial or language group. There are, of course, myriads of common or vernacular names in all countries, but in general these names are limited in use to racial and language groups, and incidentally, are often much more loosely applied than are the technical Latin names of plants.

INDIVIDUALS HONORED IN PLANT NAMES

As noted above, very numerous generic names have been derived from the names of individual men who have, in one way or another, been concerned with plant problems. Far more individuals have been honored in having species named for them than is the case with genera, because there are infinitely more different species of plants than genera. It is usually the collector who operates in inadequately explored areas who most frequently has his name perpetuated in the form of a generic or a specific cognomen. A single example will suffice. The genus *Parkia* honors the name of Mungo Park, a noted African explorer, and the binomial *Parkia roxburghii* commemorates the name of William Roxburgh, one of the pioneer botanists who worked on the flora of British India. The chances are very great that any serviceman operating in some of the out-of-the-way and little-explored parts of the world may thus have his name perpetuated if he elects to prepare and submit specimens for identification. Perhaps this might be designated as a penalty!

When the genus *Ahernia* was published in 1909, it was dedicated to the late Colonel George P. Ahern, who at that time was Director of Forestry in the Philippines. It so happened that the basal parts of its leaves were adorned with a pair of very conspicuous glands, for which reason the descriptive specific name *glandulosa* was selected, the binomial being *Ahernia glandulosa*. Shortly after the description was published, one of the foresters observed: "Why was that combination selected? It sounds like a horrible disease. Why wasn't it called Aherni O'Dendron and thus kept Irish?" This suggested name, had it been selected, would mean "Ahern's tree," so the forester had an idea. In my own case, with no less than nine generic names derived from my name, and at least 225 specific names in such forms as *merrillii* and *merrilliana*, the one that intrigues me most is *Merrilliopeltis*, applied by a German specialist to a genus of fungi, for this euphonious name means "Merrill's skin." As one mycologist expressed it: "*Merrilliopeltis* must be a pretty tough fungus!"

TECHNICAL DESCRIPTIVE WORDS

While in this volume a serious attempt has been made to avoid the use of highly technical descriptive words, sometimes their use has been unavoidable because of the lack of exact equivalents in everyday language. To assist the lay reader, a short glossary has been appended (p. 277), although the strict botanical meanings of even the more common terms appear in standard dictionaries. In all of the biological sciences, special vocabularies have been evolved, for fixed meanings are necessary when it becomes essential precisely to define the limits of an entity, whether it be family, genus, species, or other category.

Specific terms are essential to the concise and accurate descriptions of the different parts of the plant, such as the underground parts, *bulb* (as in the onion), *tuber* (as in the potato), *corm* (as in taro or our Jack-in-the-pulpit), *rhizome* (as in ginger and *Iris*), as well as the stems and the branches, and the leaves, including their arrangement, shape, type, consistency, surface and margin characters. The same thing applies to flowers and their parts, as well as to the fruits and the seeds. As to habit, such commonly used terms as herb, shrub, vine, and tree are self-explanatory, for these are used in botany as in their everyday meaning; but herbs must be differentiated as between those that live and mature their fruits within a single season, and those that live for many years. The former are characterized as *annual* and the latter as *perennial*. Naturally, all woody vines, shrubs, and trees are perennial.

LEAVES

Here some common English words are used in descriptive botany, sometimes in a specialized sense, although the majority of these descriptive terms are of Latin derivation; some of the common terms are smooth, rough, hairy, toothed, lobed, acute, heart-shaped, obtuse, and rounded, which scarcely need any further definition. But technical terms are necessary in many

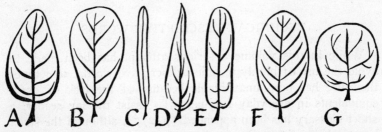

Fig. 6. LEAF SHAPES: A. ovate; B. obovate; C. linear; D. lanceolate; E. oblong; F. elliptic; G. orbicular or round.

cases, such as *glabrous* to indicate smooth leaves without any indumentum, for some hairy leaves may be smooth to the touch, *pubescent* to indicate hairy leaves, with such variants as *hirsute* when the hairs are stiff and rather short, *villose* when they are long and soft, and *tomentose* when they are soft and very densely arranged.

As to shape (fig. 6, A–G), such common words as *ovate*,

linear, lanceolate, oblong, elliptic, round (*orbicular*) indicate leaf outlines in the sense that such words are used in everyday parlance, but certain more technical names become necessary, such as *obovate*, the reverse of ovate, where the leaf is wider above the middle and narrowed below, *cordate* for heart-shaped leaves or leaf bases, and for the leaf tips *acuminate* for slender tapering tips, *acute* for sharp pointed ones, *obtuse* for blunt tips,

Fig. 7. LEAF TIPS AND BASES: A. acute; B. acuminate; C. obtuse; D. round; E. retuse; F. cordate or pear-shaped; G. peltate.

and *retuse* for notched ones. A *peltate* leaf is one where the leafstalk is attached to the lower surface, not to the margin, thus suggesting a shield (Latin *pelta*), (fig. 7, A–G).

As to consistence, a leaf may be described as *coriaceous* (leathery), when it is firm, tough, and rather thick, *membranaceous* (thin), *papyraceous* (papery), *succulent* (fleshy), and as stiff, brittle, tough, flexible, etc. As to margins (fig. 8), the word *entire* is used when there are no teeth or lobes, but it becomes necessary to define more closely certain characteristic forms of teeth, such as *dentate* when they are sharply pointed and extend outward, *serrate* when they are sharply pointed and extend upward, and *crenate* when their tips are rounded. A wavy margin, with no evident teeth, is designated as *undulate*. When the indentations are large and deep, the term *lobe* is used, and for those types where the lobes radiate from a common center, the term *palmately lobed* is descriptive.

Again, leaves, as to their position, are described as *alternate* when one is borne above another on different sides of the stem, *opposite* when two leaves are placed exactly opposite each other at a node, and *verticillate* when three or more leaves are attached at the same node, forming a distinct whorl. Further, leaves may be *sessile* (with no leafstalk), or *petioled* (with a shorter or longer leafstalk).

Leaves may be simple or compound. A *simple* leaf (fig. 8, A–G) is one with or without a leafstalk and a single blade; a *compound* leaf (fig. 8, H–J) is one where the axis bears several to many leaflets. Compound leaves are characterized as *pinnate*

Fig. 8. LEAF MARGINS AND FORMS: A. entire; B. dentate; C. serrate; D. crenate; E. undulate; F. lobed; G. a palmately lobed leaf; H. a palmately compound leaf; I. a pinnate leaf; J. a bipinnate leaf.

when the leaflets are arranged on a simple axis, as in the ash and walnut trees, the word being derived from the Latin *pinna* (feather); *bipinnate* and *tripinnate* are terms used to characterize more complex leaves, where the main axis bears secondary ones in the former, and the secondary ones bear still others in the latter, the secondary and tertiary ones bearing leaflets. Another type of compound leaf is that designated as *palmate* where several to many leaflets are attached at the apex of the leafstalk in a radiate arrangement, corresponding roughly to the arrangement of the fingers on one's hand, whence the name itself from the Latin *palma* (hand).

FLOWERS

One realizes that there are enormous differences in the shapes, sizes, colors, and other characters of flowers. They are *perfect* when the stamens and ovaries are present in the same flower, but a perfect flower is not necessarily a complete one, for the

BOTANICAL CLASSIFICATION 23

outer floral organs may be lacking. A *complete* flower (fig. 9) is one where all four sets of organs are present, the outer and usually green set, the *sepals*, together forming the *calyx*; the next inner row, usually colored, the *petals,* together forming the *corolla*, and these petals may be free from each other or more or less united; the third inner row, the *stamens*, the stalks

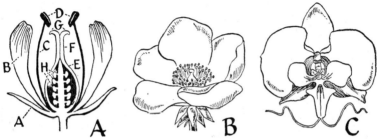

Fig. 9. FLORAL PARTS AND FLORAL TYPES: A. longitudinal section through a regular, perfect, complete flower; A. sepal (together forming the calyx); B. petal (together forming the corolla); C. filament and D. anther (together forming the stamen); E. ovary; F. style; G. stigma; H. ovules. B. a regular flower of the rose (*Rosa*), all parts of each set of organs being alike in shape and size. C. an irregular flower of an orchid (*Phalaenopsis*), the several parts of each set of organs dissimilar in shape and size.

(sometimes absent) being the *filaments* and the pollen-bearing bodies at their tips being designated as the *anthers*; the fourth or innermost organ is designated as the *ovary*, within which the ovules are produced, the latter, after fertilization by the pollen, developing into the seeds. The ovary is crowned by the sessile or stalked *stigma*, the function of which is to receive the pollen, and its stalk, if present, is called the *style*; the style and stigma together with the ovary form the *pistil*.

By no means are all flowers either perfect or complete, for in many groups of plants one or more sets of floral organs may be absent. Again, many flowers may be only *staminate* (male), or *pistillate* (female), these male and female flowers sometimes being borne on the same plant, sometimes on separate plants. Often the male and female flowers are similar in shape and size, but in some groups they are very dissimilar, as in the papaya (*Carica papaya*). Flowers may be *regular* or *irregular*; in the former all of the organs of each set are alike in shape and size as in the apple, and in the latter, such as those of the bean and the orchids, in one or more sets of organs there are distinct differences in size and shape.

Flowers may be arranged in a great variety of ways, and whatever the arrangement, taken together, they form the *inflorescence* (fig. 10, A–F). A *panicle* is a branched and rebranched, open, many-flowered inflorescence; a *spike* is a simple

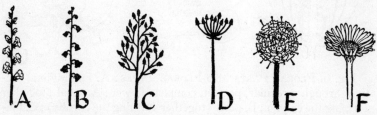

Fig. 10. Types of Inflorescences: A. spike; B. raceme; C. panicle; D. umbel; E and F. heads or *capitula*, the latter the characteristic head of the Compositae family, the numerous small flowers being surrounded by an involucre.

elongated axis bearing sessile flowers; a *raceme* is similar to a spike but the flowers are stalked; a head or *capitulum* is a dense ovoid or globose mass of numerous crowded flowers; an *umbel* is that type where the arrangement of the branches suggests the structure of an umbrella; a *fascicle* is that where few to many flowers are more or less crowded in clusters, and usually in the leaf-axils. In very many groups the flowers are *solitary*, that is, a single stalk bears a single flower.

FRUITS

The fruit is the matured ovary with its fertilized ovules developed as seeds. They vary enormously in shape, in size, and in other characters. They may be *simple* (developed from a single ovary) as in the cherry; *aggregate* (developed from several to many ovaries of a single flower) as in the strawberry and raspberry; or *collective* when they are developed from few to many separate flowers as in the mulberry (*Morus*) and the breadfruit (*Artocarpus*). In general, the fruit wall is called the *pericarp* regardless of its characters.

As most fruits have length, breadth, and relative thickness, slightly different descriptive terms are used than in describing thin or flat organs such as leaves and petals. Thus, while a round leaf might be described as *orbicular*, a round fruit would be characterized as *globose* or *spherical*. An *ovate* leaf denotes a certain shape, but a fruit having a similar outline is characterized as *ovoid*, and while a leaf that is wider above the middle and narrowed below that would be characterized as *obovate*, a fruit corresponding to this in outline would be characterized as *obovoid* or *pyriform* (pear-shaped). In fruit descriptions such descriptive terms as flattened, compressed, and cylindric are self-explanatory. Often, however, a stem, or a long, slender, more or less pencil-shaped cylindric fruit is characterized as *terete*, and a spherical fruit depressed at the end may be characterized as *depressed globose*, like some of our apples.

Fruits may be fleshy or dry, and *dehiscent* (splitting open at maturity) or *indehiscent* (not splitting open). A *berry* in botany is a fleshy fruit normally containing a great many seeds, such as the tomato; a *capsule* is a small to large, usually dry fruit, with few to many seeds, which may or may not split open regularly at maturity. A *drupe* is a fleshy fruit containing a single hard seed, as the peach and the cherry. A *pome* is that type of fruit characterized by the apple, where the fleshy part is composed of the greatly altered and enlarged calyx tube. A *pepo* is a fleshy fruit with a very large number of flattened seeds imbedded in its pulp or borne on internal sutures, being typical of the *Cucurbitaceæ* and represented by the common squash, melon, and cucumber. A *pod* or *legume* is the characteristic fruit of the *Leguminosæ* as represented by the common bean and pea, one-celled, one- to many-seeded, the seeds borne along one side of the single cell, the valves splitting along one or both sutures or remaining closed; in some groups, such as the tamarind, the mature fruits are fleshy, yet it is still a pod. Another rather common type is the *follicle* which is usually dry at maturity, splits open along one side, and contains numerous often winged seeds, or seeds provided with a distinct pappus, as in the *Apocynaceæ* and *Asclepiadaceæ*, the milkweed types. Sometimes fruits, like certain seeds, are provided with wings to aid in their distribution, as in the maple.

SEEDS

As to seeds, variation here, as with the fruits, is great, but within the limits of individual families their basic characters are often rather uniform. Frequently one can tell at a glance what natural family is represented by an individual seed, such as the squash family (*Cucurbitaceæ*), bean family (*Leguminosæ*), gutta-percha family (*Sapotaceæ*), and others. Many seeds are minute, in fact sometimes almost dust-like. Others bear thin wings or terminal tufts of soft hairs as an aid to their distribution through the agency of wind, and still others have adaptations to dissemination by floating, being carried long distances by ocean currents. Many of the eminently successful weeds have fruits or seeds that adhere closely to the fur of animals and to the clothing of man, and are thus widely spread in nature. Potent distributors of seeds are the fruit-eating birds and animals, and even migratory birds may, under some circumstances, widely extend the ranges of individual species.

3

Plants of the Seashore

In addition to those species mentioned and briefly discussed in the next chapter on the mangrove forest, a distinctly larger number are characteristic of the areas immediately back of the seashore throughout the Old World tropics. In comparison with the vegetation of the inland forested areas, this seacoast or strand flora, which is thoroughly characteristic of the habitat, is a relatively poor one, both in genera and in species. The very fact that normally the species characteristic of the strand and its immediate vicinity are seldom found any great distance from the sea, unless occasionally planted inland, makes this a relatively simple category to discuss. Again, the fact that without exception these strand plants, like those of the mangrove swamps, have fruits or seeds especially adapted to dissemination by floating in salt water, has led to an enormously wide distribution of many of the species involved; some of them, as a matter of fact, are found immediately back of the beach on all tropical seashores throughout the world. Normally, the flora of a typical low island (atoll) far out in the Pacific Ocean consists entirely or almost entirely of species whose seeds or fruits are distributed by ocean currents; and accordingly many, but not all, of the elements that will be observed along the strand in Malaysia or in the Southwestern Pacific region will be found to occur naturally, in greater or less numbers, on small, low islands throughout Micronesia and Polynesia, as well as along the seashores of the high islands in those regions.

This strand or seacoast vegetation consists of trees (some of large size), shrubs, woody vines, and herbs, and in it one notes representatives of about fifty natural families of plants. It is not plant relationships that make up this characteristic although limited type of vegetation, but rather the ability of a group of very different and botanically often not closely allied plants to thrive in a distinctly unfavorable environment in the presence of salt or brackish water. All of the species characteristic of this habitat have seeds or fruits that are distributed through the medium of ocean currents. Thus it is that we find a rather remarkably uniform strand flora all the way from East Africa and Madagascar through southern Asia, Malaysia, Micronesia, and Polynesia, to the Marquesas Islands and Hawaii.

The strand forests, where forests exist, are in general not so dense as those of the inland secondary forests or the borders of primary forests along inland streams. In the mangrove forests discussed elsewhere one will find dense stands of individual tree species, such as the true mangrove tree (*Rhizophora*), and back of certain sandy beaches usually narrow strips of *agóho* forests (*Casuarina equisetifolia*, fig. 18). This tree in form and appearance suggests the pine, but its numerous slender jointed green branchlets function as leaves and its wood is very hard. Its common name in Australia, she-oak, was apparently derived from the structure of its wood, suggesting that of the oak, and the soft soughing sound of the wind blowing through its tops. Often immediately back of the beach occur continuous thickets of the curious screw-pine or *pándan* (*Pandanus tectorius*, fig. 11, and related species). On precipitous rocky shores the inland forest sometimes extends to the very edge of the ocean. Where typical strand trees occur, other than the two above mentioned, they usually appear as scattered individuals and do not develop in pure stands.

LARGE TREES

Among the large trees in this category—and most of them are very widely distributed in the Pacific basin—are such as the *pálo maría* (*Calophyllum inophyllum*, fig. 39) of the mangosteen family always growing immediately back of sandy beaches, with its yellowish milky juice, very leathery, entire, smooth, glossy leaves, racemes of attractive white flowers, and globose, smooth, one-seeded fruits. The timber of this tree is distinctly hard and heavy and is of commercial importance. Associated with this, and almost equally large, but with relatively soft wood, is *bótong* (*Barringtonia asiatica*, fig. 41), with its much larger, obovate, entire, glossy, slightly fleshy leaves, large flowers with numerous pink stamens, and large, sharply four-angled, one-seeded fruits. These fruits have a thick fibrous wall and are distributed by floating in salt water. Incidentally, the seed is used to stupefy fish in tidal pools and in quiet streams; the firm white seed is crushed, mixed with water, and thrown into pools where fish occur. This method of stupefying fish does not in any way impair the flesh of the fish as food.

Almost equally large is another strand tree with peltate, somewhat fleshy, smooth-margined leaves, rather insignificant greenish flowers, and fruits reminding one in general of a Japanese lantern in that the single black seed or fruit proper is enveloped in a loose, free, whitish, somewhat fleshy involucre with a round opening at the top. This is *kólung-kólung* (*Hernandia ovigera* or *H. peltata*, fig. 16). Still another large strand tree is the Indian almond or *talísay* (*Terminalia catappa*, fig. 42) with fairly large obovate leaves, slender spikes of insignificant small greenish flowers, and a characteristic, somewhat compressed, two-keeled, reddish fruit about an inch long containing a single edible seed. In this species the fruit itself is usually reddish at maturity and more or less spongy in texture, the latter feature being its adaptation to dissemination through the agency of ocean currents. At maturity many of the leaves turn red before falling. The so-called Australian-pine or she-oak, *agóho* (*Casuarina equisetifolia*, fig. 18) mentioned above often forms narrow strips of gregarious forest immediately back of sandy beaches.

Not uncommon back of the beach is the *dap-dap* (*Erythrina variegata*, fig. 27), a representative of the bean family, with its dense racemes of large, bright red flowers and three-foliolate leaves, the individual flowers being about three inches long. Often conspicuous among the large trees is the *kalúmpang* (*Sterculia fœtida*, fig. 34), with palmately compound leaves and large red fruits which split open along one side; its flowers have a very disgusting odor.

SHRUBS, SMALL TREES AND WOODY VINES

In addition to the trees that may be designated as normally large at maturity, an even larger number of smaller trees and shrubs occur in this specialized habitat, as well as some woody vines. These may be summarized as follows. Perhaps the most common species is the awkward-looking *pándan* or screw-pine (*Pandanus tectorius*, fig. 11), which frequently forms dense thickets immediately back of the sandy seashore. It is a shrub or at most a small tree, bearing characteristic prop roots on the basal parts of its trunk and elongated, sharply toothed, long pointed leaves arranged in a distinct spiral at the tips of the relatively few branches. The rather large, round or ellipsoid fruits are compound, the scanty orange-red pulp surrounding the individual parts being edible, as are the rather numerous but relatively small seeds.

In some regions the very primitive *pitógo* (*Cycas circinalis*, fig. 12) grows; the family to which it belongs is a very ancient one, its scattered representatives being the most primitive of living flowering plants. It is palm-like in habit, with normally a single trunk bearing a crown of numerous leathery pinnate leaves and spiny leaf-stalks. The male flowers are borne in very large erect cones within the crown of leaves, and the female flowers and fruits are produced on separate plants on specialized narrow branches (infructescences) mixed with the leaves. The numerous smooth ovoid fruits contain a single large seed, which, when fresh, is very poisonous because it contains free hydrocyanic acid. This poisonous principle, which gives the characteristic taste to apple and peach seeds, can be eliminated by crushing the seed, soaking in several changes of water, and then drying. The end product may then be cooked and used as food. It is the *fádang* of Guam, and in times of scarcity serves as a distinctly important famine food.

Two small or medium-sized trees in the mallow family are very common. One is the *bau, fau, hau,* or *vau* of Polynesia, the *malabágo* of the Philippines (*Hibiscus tiliaceus*, fig. 37), its bark having a strong bast fiber. Its natural habitat is back of the beach but it sometimes forms dense thickets inland; the large regular flowers are yellow, turning reddish in age, while its broadly heart-shaped leaves are pale and pubescent beneath. Closely allied to this, but with ovate, very smooth leaves and also with large yellow flowers is *milo, miro,* or *banálo* (*Thespesia populnea,* fig. 38), which differs strikingly from the *Hibiscus* in that its nearly round fruits do not split open at maturity.

In the western part of the Pacific basin, also back of the seashore, but not in the mangrove swamps, is the *piágau* or *lei-lei* (*Xylocarpus moluccensis,* fig. 29). This is a tree with rough bark, somewhat pointed leaflets, and round fruits three or four inches in diameter; its large irregular seeds are distributed by floating. *Apíri* (*Dodonœa viscosa,* fig. 32) is a common shrub or small tree with narrow, smooth, somewhat sticky leaves, insignificant flowers, and thin-walled, bladdery, narrowly winged fruits about one-half inch in diameter; this, like a few other strand plants, also grows inland in favorable localities. *Molitái* or *bidáru* (*Ximenia americana,* fig. 22), the tallow-nut of Florida, is a somewhat spiny shrub with simple, entire, leathery leaves and plum-like fruits; it usually grows in wet places and, like many of the strand plants, occurs in all tropical regions. To be treated with proper respect is the very common shrub or small tree known as *búta, búta-buta, bétah,* or *alipáta* (*Excœcaria agallocha,* fig. 31) because its abundant milky sap is an irritant, and especially dangerous if brought in contact with mucous membranes such as the tissues surrounding the eyes; under such circumstances it causes intense inflammation and has even been accused of causing blindness. Its small insignificant flowers are borne in spikes, while the firm smooth leaves are entire.

Tábau or *terúntum* (*Lumnitzera littorea,* fig. 43) is recognized by its rough bark and brilliant red flowers borne in terminal inflorescences, its entire, firm smooth leaves being narrowed below; and occasionally one will note the closely allied small shrub with white flowers, *Lumnitzera racemosa*. The characteristic species of the Rhizophoraceæ, the true mangrove swamp trees, and certain others especially typical of the mangrove areas, are discussed in the chapter on the mangrove forest; occasionally one observes scattered representatives of these along the strand in favorable localities outside the swamp areas. Another shrub or small tree, frequently very abundant, and universally distributed in the Old World tropics, is *bantígi, mentígi,* or *sentígi* (*Pemphis acidula*, fig. 36), which may be distinguished at once by its small regular white flowers and its numerous small leaves with a distinctly acid flavor. *Bálok-bálok, mempári, bangkóng* (*Pongamia pinnata*, fig. 19), *tambalísa, pelótok* (*Sophora tomentosa*, fig. 28), and *malapígas, blának* (*Desmodium umbellatum*, fig. 21) are shrubs or small trees representing the bean family in the habitat under discussion. All are common. The first has rather small white flowers and hard, smooth, one-seeded pods and three large smooth leaflets; the second has densely and softly pubescent, pinnate leaves, with many leaflets, yellow flowers, and elongated cylindric pods distinctly constricted between the seeds; while the third has small white flowers, medium-sized, three-foliolate leaves, and small, distinctly jointed pods.

Pilápil (*Ægiceras corniculatum*, fig. 61, and *A. floridum*, fig. 62), mentioned also in the discussion of the mangrove vegetation, is a shrub with thickened, entire, obovate leaves, small pink regular flowers borne in terminal umbels, and small, distinctive, curved, horn-shaped fruits. *Lipáta* or *bintáru* (*Cerbera manghas*, fig. 40) may be recognized by its rather large, smooth, elongated leaves, its fairly large, regular, white flowers with a purple eye, its abundant latex, and large, purple, generally ellipsoid, smooth fruits. The borage family is represented by two characteristic species. One is *agútug* or *murmasáda* (*Cordia subcordata*, fig. 45), a small tree with large smooth leaves, fairly large reddish flowers, and small globose fruits; the other, occurring only on sandy beaches and not in the inland thickets, is *salakáho* or *babakóan* (*Tournefortia argentea*, fig. 47), a low shrub with large, pale, densely pubescent leaves, its numerous small flowers in arrangement reminding one of the heliotrope. Almost everywhere on the sandy beaches *lagúndi* (*Vitex trifolia* var. *simplicifolia*, fig. 46) will be noted, a prostrate, widely creeping vine, with irregular blue flowers, small fruits, and round to obovate, simple leaves; branches of this may extend for long distances over the loose sand. In inland thickets and back of the beach one may also observe the erect form of *lagúndi* (*Vitex trifolia*), the leaves usually with three narrow, sharp pointed leaflets, quite different from those of the prostrate form that thrives on the sandy beach; in both forms the leaves are rather pale beneath.

Saminánga (*Clerodendron inerme*), an erect shrub with opposite entire leaves and slender, white to purplish, elongated flowers, is frequently abundant; and everywhere on the broad sandy beaches above high-tide limits will be noted the singular *bóto* (*Scævola frutescens*, fig. 48), a low erect shrub with few branches, large, obovate, somewhat fleshy, entire leaves, pale blue very irregular flowers, and small, somewhat succulent, white fruits. In the coffee or madder family one common representative will be *bankúdo* or *nino* (*Morinda citrifolia*, fig. 49), a shrub or small tree with large, entire, glossy, smooth, opposite leaves, the tubular regular white flowers scattered here and there on the short inflorescence which eventually forms a fleshy, ovoid, white fruit. Another common representative of this family, also characteristic of the strand, is *búa-búa* or *tabúg* (*Guettarda speciosa*, fig. 50), also a shrub or small tree, with distinctly large, more or less pubescent leaves, tubular white flowers, and rather small, hard, smooth, brown, depressed-globose fruits.

Oúru or *ugágoi* (*Suriana maritima*, fig. 23) is found only in the open sandy places and chiefly on small islands; a small erect shrub, it may be recognized by its small yellow flowers and its small, narrow, somewhat fleshy, entire leaves densely crowded on the branchlets. In this assemblage of strand plants, *kabatíti* (*Colubrina asiatica*, fig. 35) is unusual because of its heart-shaped, rather thin and somewhat crenately toothed leaves; this scrambling shrub, which belongs in the buckthorn family (*Rhamnaceæ*) has small greenish flowers and small obovoid black fruits. Particularly in the western parts of the area, one or two species of *Allophylus* will be noted, sometimes known as *handámo*; these sprawling or semi-scandent shrubs have trifoliolate, irregularly sinuate, toothed or nearly entire, fairly large leaflets, small greenish white flowers in lax panicles or racemes, and small globose red fruits. One species of *Jasminum*, sometimes known as *manól*, a sprawling woody vine with white regular flowers (*J. bifarium*, fig. 63) is frequently abundant, particularly in wet brackish places. Occasionally one will note the curious *gógo* (*Entada phaseoloides*), a coarse woody vine characterized by its enormous pods and very large brown seeds. The crushed stem is used as a soap substitute.

GRASSES AND SEDGES

In addition to the woody species practically restricted to the strand areas are also a number of grasses and sedges which are not considered in detail here. However, the very coarse and widely creeping *tikúsan* (*Spinifex littoreus*, fig. 13) with its globose heads of crowded flowers bearing long slender spines, is worthy of mention among the grasses. This is a Malaysian plant, not extending to Polynesia. Another interesting one is the semi-prostrate *rúmpet kerúpet* (*Ischæmum muticum*) with its inflorescences once-forked, the forks being closely appressed. Perhaps a third might be included, *Thuarea involuta*, a creeping slender grass which, because of its very peculiar fruits, one botanist named *Ornithocephalochloa*, that is, birds-head grass. Various sedges will be observed, in such genera as *Cyperus, Mariscus, Eleocharis, Fimbristylis,* and *Scleria,* and the low rigid *Remirea maritima.* The latter, growing on open sandy beaches, is a plant of world-wide distribution in the tropics, and always in this habitat. Generally speaking, most tropical sedges and grasses are of very wide geographic distribution.

PALMS

In the palm family, little is to be noted, for the ubiquitous coconut palm (*Cocos nucifera*) normally occurs only where it has been planted by man. In the Malay Archipelago proper, other than the swamp-loving nipa palm (*Nipa fruticans*) discussed in the chapter on mangrove forests, there are very few distinctly halophytic palms, i.e., adapted to growth in places where salt or brackish conditions prevail; but the nipa palm scarcely extends beyond the Southwest Pacific area. Not uncommon in Malaysia back of the seashore, is the small palm with very deeply divided, fan-shaped leaves with strong leafstalks, *úgsang* or *balatbát* (*Licuala spinosa*).

MONOCOTYLEDONOUS HERBS

Among the herbaceous plants adapted to this specialized habitat is one representative of the amaryllis family, *bákong* (*Crinum asiaticum*, fig. 14), a coarse erect plant with broad elongated leaves, often with a distinct stout but short trunk, large, regular, white, fragrant flowers with six long narrow segments, and fairly large fruits. It prefers the sandy soil immediately inland from the beach line. In an allied family is the distinctive Polynesian arrowroot *pía* or *masóa* (*Tacca pinnatifida*, fig. 15), with its tall, erect, fluted green stems often three feet high, thrice-divided and strikingly lobed leaves, and peculiar flowers borne on separate leafless tall stalks; its hard potato-like tubers, produced in the loose sand, are rich in starch, but are very bitter and poisonous unless properly treated by crushing, thorough washing, drying, and cooking. After being thus processed, they may be used as food. Its natural habitat is in the loose sand well above the limits of normal high tide, the plants preferring partial shade. Among the few other monocotyledenous groups that have representatives in the strand flora is the vine *láfo, turúka,* or *úag* (*Flagellaria indica*), which can always be recognized by the tips of its narrow smooth leaves being transformed into tendrils for climbing; naturally this does not occur on the open beach but rather in thickets back of the seashore.

DICOTYLEDONOUS HERBS

Among the dicotyledonous herbs, *katúri* or *ríma* (*Boerhavia diffusa*, fig. 17) may be noted everywhere back of the seashore and even on small coral islands where vegetation is at all developed; this is a rather diffuse herb with smooth, somewhat fleshy leaves and very small pink flowers. Associated with it, and one of the very commonest and most widely distributed of all strand plants, is the slender yellow or yellowish green leafless parasite, *táli-putéri, aírie,* or *aímoa* (*Cassytha filiformis,* fig. 24), bearing small globose or ovoid fruits. This plant is parasitic on a great variety of other species. Although simulating the common dodder in appearance, it is totally unrelated to the latter, belonging instead to the laurel family. *Dampálit* or *gélang-laut* (*laut* = ocean) (*Sesuvium portulacastrum,* fig. 20) is a prostrate herb with very fleshy narrow leaves and small flowers and fruits. This succulent plant may be eaten raw or cooked, but it usually contains so much salt that it is desirable to soak or parboil it in fresh water to remove a part of the salt before eating. In a closely related family is the common purslane, *golasíman* or *gélang* (*Portulaca oleracea,* fig. 122), a very succulent plant with fleshy, usually prostrate stems, thick, smooth leaves, and regular yellow flowers. This and allied species also may be eaten either cooked or raw.

Two representatives of the linden family are characteristic of this particular habitat, both being members of the large tropical genus *Triumfetta,* a group distinguished by their yellow flowers and small, round, prickly fruits, the prickles usually ending in small hooks. Both of the strand species are prostrate creeping herbs, the most common and widely distributed of the two being *kubíli* or *titíti* (*Triumfetta procumbens,* fig. 30), coarser, with distinctly pubescent, somewhat undulate but scarcely lobed leaves, and with larger flowers and fruits than the less common *Triumfetta repens,* which has smaller, distinctly three-lobed leaves. And not to be overlooked in the herbaceous group is the erect, sparingly branched, glabrous *aatási, atóto,* or *tóto* (*Euphorbia atoto,* fig. 33), with an abundance of white latex, everywhere on sandy seashores.

VINES ON WIDE SANDY BEACHES

On the broad sandy beaches, often spreading for great distances, are certain herbaceous vines, the most characteristic being *patáning dágat* (*dágat* = ocean), *pítai, pirítai* (*Canavalia maritima,* fig. 25) and *nanéa, akánkan* (*Vigna marina,* fig. 26) of the bean family, and the beach morning-glory, *tagárai, fúe, poúe, pohúehúe* (*Ipomœa pes-capræ,* fig. 44). The *Canavalia* has attractive pink flowers, while an allied species with broader pods and somewhat pointed leaves (*C. microcarpa,* fig. 25A) occurs in the thickets back of the beach. The *Vigna* has small, nearly cylindric pods and yellow flowers. The most striking plant in this assemblage is the beach morning-glory, its leathery leaves having two broad rounded lobes suggesting the shape of a goat's foot, whence the specific name *pes-capræ*=goat's foot; the attractive flowers are pink to pale purple. Within this category are also the widely creeping, very coarse grass, *Spinifex littoreus,* fig. 13, mentioned above under grasses and sedges, and the woody but widely spreading *Vitex trifolia var. simplicifolia,* fig. 46, with its irregular small blue flowers, briefly discussed under shrubs, small trees, and woody vines.

In the thickets back of the beach one may observe certain vines representing the milkweed groups, Apocynaceæ and Asclepiadaceæ, also some semi-epiphytic representatives of the genera *Hoya* and *Dischidia* of the latter family, with their thick fleshy leaves and their abundant latex. And often very abundant in the beach thickets are certain members of the Compositæ, such as the somewhat climbing *áte-áte* or *hagónoi* (*Wedelia biflora,* fig. 51) with its yellow flowers and the small erect shrubby *kalapíni* (*Pluchea indica*) with its small purplish flowers. There is also commonly one representative of the squash family, the wild form of the loofah or dishcloth gourd, *patóla, tabóbok,* or *motíni* (*Luffa cylindrica,* fig. 212A), with yellow flowers and fairly large, cylindric, smooth fruits. Often plentiful will be various species of the morning-glory family (Convolvulaceæ), particularly the moonflower *terúlak* (*Calonyction album*), rampant in thickets, with its great white flowers open only at night. Characteristic also of the strand is the rather delicate *Ipomœa gracilis,* with its ovate smooth leaves and pale purple flowers about one and one-half inches in diameter.

In this brief summary of the dominant species that will be observed on and immediately back of the strand and generally throughout the tropical parts of the Pacific basin, no consideration has been given to intruders in the form of more or less ubiquitous weeds, for these properly are not parts of the original strand vegetation. Most of them happen to be accidental introductions into these regions from other parts of the world within the historical period; the number of these intruders in the beach vegetation is small. One of these is perhaps worthy of note because wherever it has been introduced it is so perfectly at home on the strand. This is the American *Catharanthus roseus* of the Apocynaceæ, a small erect plant with milky sap, opposite, oblong leaves, and showy regular flowers one and one-half to two inches in diameter, which vary from white to pink or red, or white with a pink or red center. A separate chapter is devoted to the weeds and weed-like plants. In passing, it is perhaps well to note that most of the plant species adapted to the conditions that prevail on and back of the seashore, have simple, very leathery or more or less fleshy leaves, with entire rather than toothed margins, although some thin-leaved and some pubescent-leaved species do occur. This applies in general to trees, shrubs, vines, and herbs. Clearly all plants that thrive in this habitat must have vegetative and other parts that will not be burned or injured by salt spray.

SEEDS AND FRUITS ON THE STRAND

In many places ocean-distributed fruits and seeds of many of the typical strand and mangrove plants will be observed along the seashore, frequently far distant from their places of origin in the flotsam and jetsam cast ashore here and there. It will naturally be the larger and medium-sized ones that will be conspicuous, such as the fruits of the coconut, *Barringtonia, Heritiera, Pandanus, Rhizophora, Bruguiera,* and such large seeds or fairly large seeds as those of *Xylocarpus, Entada, Pangium, Canavalia,* and others. But it should be remembered that in many species even the small or minute seeds or fruits of strand and mangrove species have their special adaptations to distribution through the medium of ocean currents, for this is the reason that the majority of such species are of such very wide geographic distribution. The seeds or fruits of these coastal species will remain afloat for many months, the seeds still retaining their ability to germinate. Frequently when cast ashore and coming in contact with moist soil and fresh water, the seeds readily germinate. Yet there are distinct limitations to this method of plant distribution, for many common Malaysian strand plants failed to reach Polynesia, and after all, there are very few of these species that occur naturally in both hemispheres, such as *Hibiscus tiliaceus, Thespesia populnea, Ximenia americana, Vigna marina, Entada scandens, Canavalia maritima, Cæsalpinia crista, Sesuvium portulacastrum, Cassytha filiformis,* the beach morning-glory *Ipomœa pes-capræ, Remirea maritima, Suriana maritima,* with a few others.

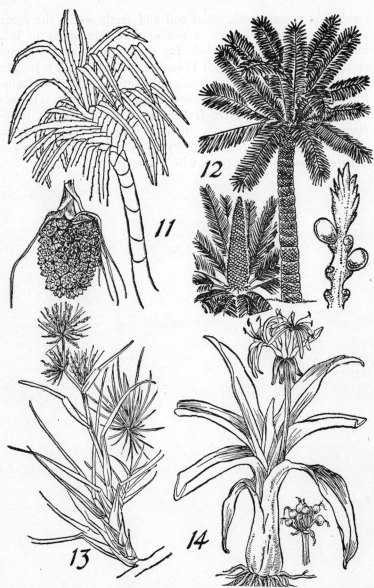

STRAND PLANTS: Fig. 11. *Pandanus tectorius*; Fig. 12. *Cycas circinalis*; Fig. 13. *Spinifex littoreus*; Fig. 14. *Crinum asiaticum*.

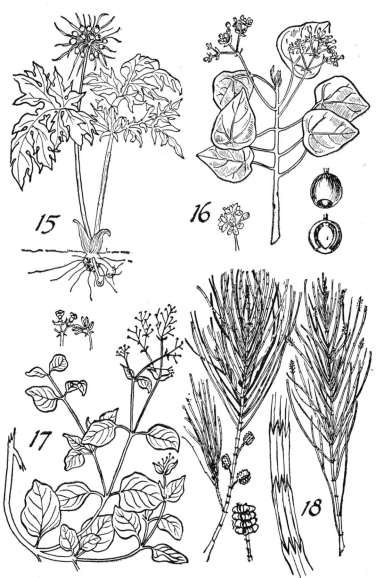

STRAND PLANTS: Fig. 15. *Tacca pinnatifida*; Fig. 16. *Hernandia ovigera*; Fig. 17. *Boerhavia diffusa*; Fig. 18. *Casuarina equisetifolia*.

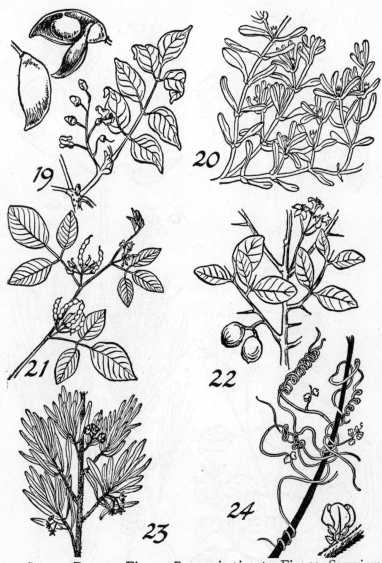

STRAND PLANTS: Fig. 19. *Pongamia pinnata*; Fig. 20. *Sesuvium portulacastrum*; Fig. 21. *Desmodium umbellatum*; Fig. 22. *Ximenia americana*; Fig. 23. *Suriana maritima*; Fig. 24. *Cassytha filiformis*.

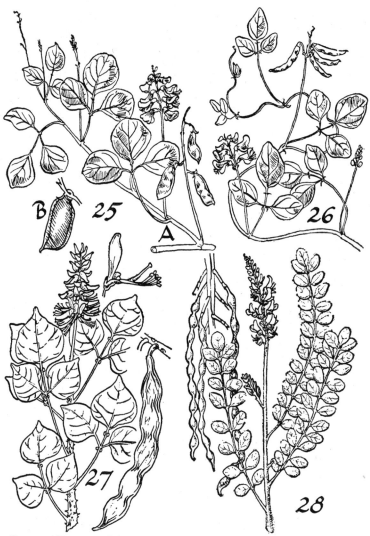

Strand Plants: Fig. 25. *Canavalia*, A. *C. maritima*, B. *C. microcarpa*; Fig. 26. *Vigna marina*; Fig. 27. *Erythrina variegata*; Fig. 28. *Sophora tomentosa*.

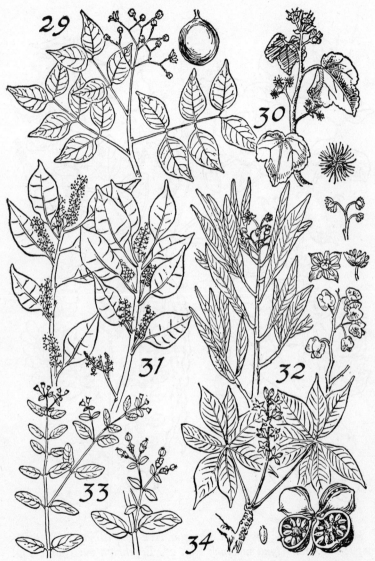

STRAND PLANTS: Fig. 29. *Xylocarpus moluccensis*; Fig. 30. *Triumfetta procumbens*; Fig. 31. *Excoecaria agallocha*; Fig. 32. *Dodonaea viscosa*; Fig. 33. *Euphorbia atoto*; Fig. 34. *Sterculia foetida*.

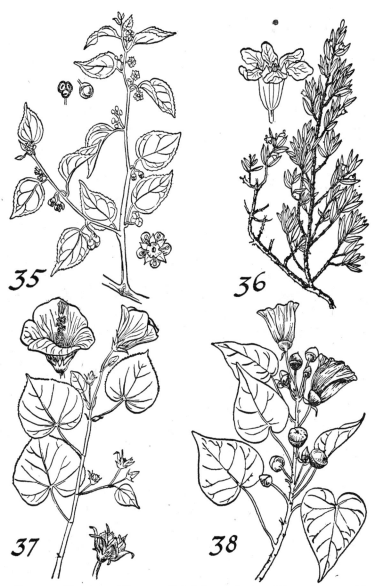

STRAND PLANTS: Fig. 35. *Colubrina asiatica*; Fig. 36. *Pemphis acidula*; Fig. 37. *Hibiscus tilaceus*; Fig. 38. *Thespesia populnea*.

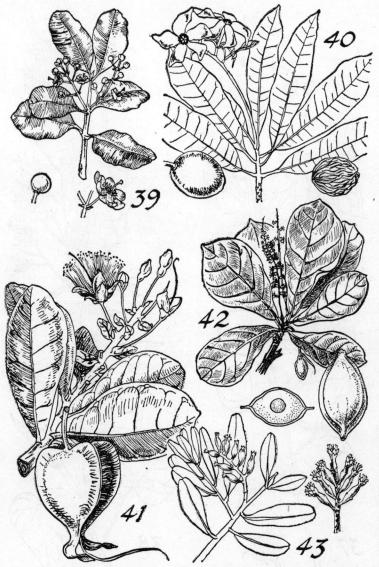

STRAND PLANTS: Fig. 39. *Calophyllum inophyllum*; Fig. 40. *Cerbera manghas*; Fig. 41. *Barringtonia asiatica*; Fig. 42. *Terminalia catappa*; Fig. 43. *Lumnitzera littorea*.

Strand Plants: Fig. 44. *Ipomoea pes-caprae*; Fig. 45. *Cordia subcordata*; Fig. 46. *Vitex trifolia* var. *simplicifolia*; Fig. 47. *Tournefortia argentea*.

STRAND PLANTS: Fig. 48. *Scaevola frutescens*; Fig. 49. *Morinda citrifolia*; Fig. 50. *Guettarda speciosa*; Fig. 51. *Wedelia biflora*.

4

The Mangrove Forest

In the tropics of both hemispheres, but most highly developed in the Malaysian region, a very characteristic type of vegetation is that of the mangrove swamp forests. This type of forest, totally foreign to temperate regions, consists of trees actually growing in salt or brackish water below the limits of high tide, the trunks of the dominant species being supported by prop roots. In addition, a number of aerial roots extend downward from the main branches of the trees. This strange forest is a phenomenon that always attracts the attention of the individual during his first visit to the tropics.

THE TRUE MANGROVE TREES

The characteristic genus in both the Old and the New World is *Rhizophora,* the name itself meaning "root bearer." These mangrove forests are of the gregarious type; that is, they are wholly or largely composed of a single species or a few very closely allied ones. This is in striking contrast to the ordinary type of tropical forest, which is composed of a great number of entirely unrelated trees. This gregarious type of forest is sufficiently developed in temperate regions, as illustrated by the vast stretches of evergreen forests consisting of pine, spruce, hemlock, and fir, but is definitely not characteristic of the tropics where the gregarious type of forest is limited chiefly to the mangrove.

These dense mangrove forests are very gloomy, for aside from the typical absence of other kinds of trees than the mangrove (*Rhizophora*) itself, except toward the inner margin, animals are absent, and few birds and few insects other than the mosquito will normally be seen. There is nothing to diversify the monotony of this depressing scenery, and even the professional botanist or collector will find little to attract and hold his attention, other than the strange environment. Another factor impressing the visitor is the relative quietness of the scene; about the only sounds that will be heard are those caused by the snapping together of bivalve shells, for shellfish attach themselves to the pendent tree roots. There are also the less obvious sounds made by crustaceans, or the splash of a fish in the channel, for a certain amount of marine life is always associated with the mangrove forests.

The mangrove forests are most highly developed in those areas covered by salt or brackish water at high tide, such as the broad mud flats along the lower reaches of tidal streams, and the borders of lagoons and estuaries more or less protected against heavy wave action and somewhat sheltered from high winds. Wherever favorable habitats occur between the Tropic of Cancer and the Tropic of Capricorn, typical mangrove forests develop, and they extend to the eastward in the Pacific basin through Micronesia as far as Fiji, Samoa, and Tonga. It is, however, on the larger islands in the Western and Southwestern Pacific that the greatest development is noted. These characteristic forests may consist of a mere fringe of trees, or where wide mud flats occur, covered by shallow water at high tide, they may be a mile or more in width. In such places where the mangrove forest is traversed by stream beds, deep penetration is possible by boat. They never extend inland beyond the reach of brackish water.

The mangrove forest is relatively dense, for the trees stand fairly close together. At maturity the trees are of a uniform height, and the thick glossy evergreen leaves form a rather dense canopy. Normally one notes that in vast stretches of mangrove forests only a single tree species will be present, and this a *Rhizophora*; although along the margins, especially on the land side, secondary mangrove forest types occur in greater or lesser abundance. There is practically no undergrowth, except on the land side where the terrain commences to rise above the level of the swamp. Because of the very extensive development of prop roots, these forests are particularly difficult to traverse, as one

THE MANGROVE FOREST

must step from slippery root to root, these often being set at dangerously sharp angles, so that each misstep plunges one into the deep soft mud.

Dominating the Oriental mangrove forests are two species of *Rhizophora* rather widely known as *bakáu* or *bakáuan* (*R. apiculata,* fig. 52A, and *R. mucronata,* fig. 52B), but associated with these are several species of the allied *busáin* (*Bruguiera conjugata,* fig. 53, and *B. sexangula*). These also have prop

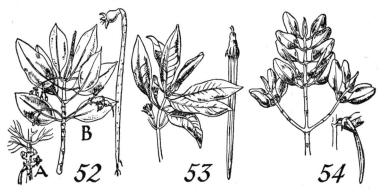

Typical Mangrove Trees: Fig. 52. *Rhizophora,* A. *R. apiculata,* B. *R. mucronata;* Fig. 53. *Bruguiera conjugata;* Fig. 54. *Ceriops tagal.*

roots. Two other species of *Bruguiera* with smaller flowers, fruits, and leaves occur, *B. cylindrica* and *B. parviflora,* as well as representatives of still another genus in the same family, *tangál* (*Ceriops tagal,* fig. 54); the latter is smaller and without prop roots. In parts of Malaysia still another genus of this family is present (*Kandelia*).

SECONDARY MANGROVE TREES

Toward the land side of the mangrove swamp, aside from the secondary representatives of the family Rhizophoraceæ, other tree species appear. Like the true mangrove trees, these are adapted to growth only along and immediately back of the seashore and within the influence of salt or brackish water, and some actually are confined to brackish marshes. Among these trees are the very large, smooth-barked *tabígi* or *piágau* (*Xylocarpus granatum*, fig. 56), with its large globose fruits six inches in diameter, resembling cannon balls, with few large irregular corky seeds; *pagátpat* (*Sonneratia caseolaris*, fig. 60), with its obovate fleshy leaves, peculiar fruits, and characteristic "breathing roots" extending upward through the mud and water and thus supplying the submerged tree roots with air; *dúngon* or *dúngon-láte* (*Heritiera littoralis*, fig. 57), with its leaves silvery beneath, and its peculiar, keeled, boat-shaped, brown fruits; and above all, the so-called white mangrove, *ápi-ápi* (*Avicennia marina*, fig. 55), with its grayish leaves, rather softly pubescent beneath, and small insignificant flowers. Here also one may note a representative of the coffee family (*Scyphiphora hydrophyllacea*) with opposite leaves and small axillary white flowers, a small tree known in the Philippines as *nílar* or *nílad* and presumably the name from which the word Manila was derived. The prefix *ma* means "the place of" or "where is"; if this explanation be correct, then the name Manila means "the place where the *nílar* grows." Along the smaller inland watercourses not infrequently one notes *túe* or *káyu-járau* (*Dolichandrone spathacea*), a tree with pinnate leaves, large leaflets, very long, slender, tubular white flowers, and long capsules which contain numerous rectangular, thin, somewhat winged seeds.

TERRESTRIAL FERNS

Normally one does not consider that salt or brackish marsh conditions are favorable for the development of ferns. Yet under these conditions one finds two very characteristic species, both widely distributed. In such places, immediately back of the mangrove forests and along and in shallow brackish swamps, occurs the very abundant *lagólo*, *páku laut* (*páku* = fern and *laut* = ocean), *péye* or *kerákas* (*Acrostichum aureum*, fig. 66A). This is a very coarse tufted fern four to six feet high, with thick leathery leaves, the fertile leaflets being continuously covered on the back with spore cases. The other is the rather coarse climbing species, *dilíman*, *hagnáya*, *lembíding*, or *pákis bang* (*Steno-

chlæna palustris, fig. 66B), the pinnate sterile fronds with broad leaflets, the uppermost fertile ones with narrow leaflets. The young parts of this fern are eaten either raw or cooked. Its tough stems are widely used for tying together the parts of fish traps or other structures subject to the influence of salt water.

TREES, SHRUBS, VINES, AND OTHER PLANTS

Immediately back of the mangrove forest, and between it and the upland forests, one notes a limited number of characteristic plants including small trees, shrubs, vines, and a few herbs. Among these are certain grasses and sedges in the more open places, and some of the strand plants more particularly dominant back of the open beaches. A small woody vine, *túba laut* (*Derris trifoliata,* fig. 59), with its thin pods and small white flowers resembling those of the common bean, is frequently abundant and may be gregarious over considerable areas. Along and in shallow pools and on the margins of watercourses with the *Derris,* the spiny-leaved *diliúario* or *doloáriu* (*Acanthus ilicifolius,* fig. 64) is prominent, its stiff smooth shining leaves, although much larger, reminding one of holly leaves; its rather showy, very irregular flowers vary from white to pale blue. Among the woody vines are the more or less spiny representatives of *Cæsalpinia.* All these have yellow flowers; one with small flat unarmed pods is *sapínit* (*Cæsalpinia nuga,* fig. 58); another larger form, *kalumbibít* (*C. crista*) has numerous leaflets in the compound leaves, very spiny pods, and large, grayish, hard, marble-like seeds.

Bagnít (*Tristellateia australasiæ,* fig. 65), a woody vine with numerous small yellow flowers and characteristic small fruits made up of three separate, somewhat star-shaped parts, each with spreading narrow wings, is widely distributed. Often a jasmine is present in abundance, a small woody vine with opposite leaves and attractive white flowers; this is *Jasminum bifarium,* fig. 63. *Desmodium umbellatum,* fig. 21, is usually common; it is a small shrub or tree, its leaves having three leaflets and its fruits being small jointed pods. Another shrub with leathery, obovate leaves, small pink flowers in simple terminal umbels, and characteristic small, somewhat curved, horn-shaped fruits is *pilápil* (*Ægiceras corniculatum,* fig. 61); the very closely allied *A. floridum,* fig. 62, has somewhat smaller leaves and compound inflorescences. The shrub or tree with rather narrow leaves strongly tapered below and with racemes of bright red flowers is *tábau* or *terúmtum* (*Lumnitzera littorea,* fig. 43).

One will further note such shrubs and small trees as *Excœcaria agallocha,* fig. 31, with its small insignificant flowers and its abundant but unfortunately somewhat caustic milky sap; *Pemphis acidula,* fig. 36, with its small white flowers and small leaves; and even occasionally the seacoast mallow, *Hibiscus tiliaceus,* fig. 37, as well as the allied *Thespesia populnea,* fig. 38. These, however, are mostly more characteristic of the open coasts, the last two with large, regular, yellow flowers. The rather awkward-looking screw-pine or *pándan* (*Pandanus tectorius,* fig. 11), is another plant also more characteristic of the open coast. In some places the nipa palm (*Nipa fruticans*) exclusively occupies fairly large areas. This is a trunkless palm with long pinnate leaves, its large, globose, dark brown, compound fruits as large as one's head being borne on special stalks arising from the underground stems. This is one of the very few palms that is adapted to growth within the influence of salt water, and wherever it grows, its leaves are utilized for thatch by the natives in preference to other material.

EPIPHYTIC PLANTS

Epiphytic vegetation in the mangrove forests may be entirely absent or, in regions of well-distributed rainfall, may be fairly well developed, although the number of different species is very limited as compared with what one observes in the higher-altitude, primary inland forests. Epiphytes are plants that grow on trees but take no nourishment from their supporting hosts, being literally "air plants." There are almost no mosses or liverworts or even lichens. The orchids are few in number and are mostly small insignificant plants, although a few large species may be seen. Some epiphytic ferns occur, particularly *pákpak laúin* (*Drynaria quercifolia,* fig. 110), with its large fertile fronds and its greatly reduced sterile ones; the latter somewhat remind one of an oak leaf in shape, and they persist as brown bract-like structures long after they have ceased to function as leaves, forming a series of bracts adapted to holding decaying vegetable debris and retaining moisture for the needs of the plant.

In the wetter regions one may even find the very characteristic bird's-nest-fern (*Asplenium nidus*) growing on the branches or in the forks of mangrove trees, especially toward the inland margins of the swamp. Also at times there is still another strange fern (*Polypodium sinuosum*, fig. 94), its fronds narrow and entire, but its thickened scaly rootstocks creeping along the branches, which are hollow and inhabited by myriads of small black ants. Perhaps the most interesting epiphyte that will be noted is *Myrmecodia tuberosa,* fig. 96, with its greatly enlarged basal parts honeycombed with large passages and these passages also occupied by myriads of small ants. Thus we may find two striking examples of symbiosis within the limits of a mangrove forest, between living plants and ants, for this symbiosis (the living together of plants and animals) is manifestly to the advantage of both the plant and the ant. The former secures a part of its nutrient requirements from the ants, and the plant itself provides shelter for the insects.

A STRANGE BIOLOGICAL PHENOMENON

Biologically, the mangrove forest is a remarkable phenomenon, but of interest chiefly as a type of vegetation wherein its component species have been developed and adapted to conditions totally at variance with the general concept as to what is necessary to plant growth. The adaptation of the primary and the secondary species of the mangrove forests to what offhand may impress one as impossible conditions for the growth of any living plant is manifest, for the removal of a plant from its special habitat results in its immediate death.

To these remarkable conditions, that is, the adaptation of many different kinds of plants to growth within the influence of salt or brackish water, among the trees and shrubs alone we find representatives of about a dozen quite unrelated families. And regardless of the family to which a given species may belong, the vegetative adaptations are remarkably alike, for most of the leaves are not only entire rather than toothed, but they are also leathery in consistency and usually smooth and glossy, although in the white mangrove (*Avicennia*) the leaves have a protecting covering of hairs.

Perhaps the most striking biological adaptation to environment, other than the fact that these true mangrove trees will grow only within the limits of salt or brackish water, is that of the seeds. In the Rhizophoraceæ, the solitary seeds in the relatively small fruits germinate while the fruit is attached to the tree and produces a long thick pendent radicle (primary root). In *Rhizophora* and some species of *Bruguiera* this radicle may be up to a foot or one and one-half feet long before the germinated seed falls from the tree, but in other species of *Bruguiera* and in *Ceriops* it may be only a few inches long and also much more slender. As these actually fall, the thickened heavy radicle may plunge into the soft mud, and a young seedling commences to develop at once. More frequently they float, sometimes for relatively great distances, and on reaching a favorable habitat strike root in the mud. Sometimes in shallow protected waters distant from the land, one notices young mangrove trees commencing their growth far distant from the forests whence their peculiar fruits came.

From the standpoint of floristics, these often vast mangrove swamp forests are most disappointing because of the very limited number of species that will be found within their confines, and again because invariably all species of plants noted in and near mangrove forests will be found to be of very wide geographic distribution. However, this wide distribution is but natural, for in every species the seeds or the fruits are adapted to dissemination by ocean currents.

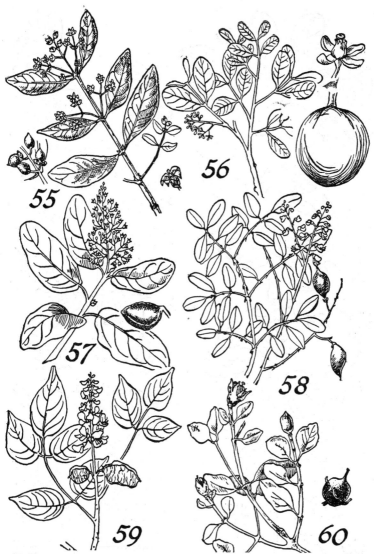

Secondary Mangrove Forest Types: Fig. 55. *Avicennia marina*; Fig. 56. *Xylocarpus granatum*; Fig. 57. *Heritiera littoralis*; Fig. 58. *Caesalpinia nuga*; Fig. 59. *Derris trifoliata*; Fig. 60. *Sonneratia caseolaris*.

SECONDARY MANGROVE FOREST TYPES: Fig. 61. *Aegiceras corniculatum*; Fig. 62. *Aegiceras floridum*; Fig. 63. *Jasminum bifarium*; Fig. 64. *Acanthus ilicifolius*; Fig. 65. *Tristellateia australasiae*; Fig. 66. A. *Acrostichum aureum*, B. *Stenochlaena palustris*.

5

The Secondary Forests and Open Grass-Lands

Everywhere in Malaysia where man has long exploited the soil, vast areas occur that are covered with a very characteristic secondary forest. The constituent genera and species in these jungle areas are very different from those that typify the primary or undisturbed forests. Man is the agent, for he provided the habitat through his destruction of the original primary forest, but nature took over immediately he abandoned this or that area that had been cleared for agricultural development, and soon the earth was again covered with characteristic, small, rapidly growing trees. This same type of vegetation also is developed in certain parts of the larger islands in the Pacific basin, such as Fiji and Samoa.

PRIMITIVE AGRICULTURE

A very common type of primitive agriculture is the simple expedient of felling the trees on the area selected for development, burning the felled material, and then planting a crop of yams, sweet potatoes, upland rice, maize, cassava, taro, or some other food-producing crop. Under primitive conditions, such quickly cleared lands are usually cultivated for only a few years at most, often for only a year or two. Erosion, loss of fertility of the soil, or invasion of the cleared area by coarse grasses, bamboo, or a mixture of quickly growing shrubs and small trees, forces the primitive agriculturist to select another forested area and repeat his operations. Once abandoned, a deserted clearing reverts very quickly to one of three types of vegetation—coarse grasses with highly developed underground rootstocks with which the primitive agriculturist cannot compete, dense thickets of bamboo, or dense forests made up of very rapidly growing shrubs and small trees. If the coarse grasses occupy the land, the area is rather permanently out of agriculture and rarely becomes reforested. If bamboo or the dense mixed secondary forest type develops, then after a few years the area may again be cleared and utilized for agricultural purposes for a short time.

This primitive agriculture is notoriously destructive of primary forests, for such a forest once destroyed is never directly replaced by the same type of vegetation. Perhaps if the devastated area remained undisturbed for a century or more, the primary forest would in time re-establish itself; but usually the pressure of population, and in open areas the annually recurring grass fires during the dry season, effectively inhibit natural reforestation with the types of trees characteristic of dense tropical primary forests.

DIFFERENCES BETWEEN PRIMARY AND SECONDARY FORESTS

The secondary forest, sometimes covering extensive areas, is totally different from the primary forest in species and to a distinctly high degree also in genera. It is this type of forest, rather than the primary type, that the word "jungle" connotes. The older secondary forests are composed of a great variety of small trees and shrubs, often also with a considerable number of vines. Many of the trees have relatively short trunks and these often branch freely like shrubs, the result being that the ordinary secondary forest is distinctly more difficult to traverse than is the primary type. In these forests, depending more or less on the density of the stand, there may be more herbaceous species and grasses than are normally found in the undisturbed primary forest, particularly if more or less open places occur.

RECENTLY DESERTED CLEARINGS

In recently deserted clearings one frequently finds that the great mass of the rapidly growing shrubs and small trees belong in relatively few genera. Taking the lead are such plants as the *hanagdóng* (*Trema orientalis*, fig. 71), belonging in the elm

THE SECONDARY FORESTS 61

family, its leaves often white-hairy beneath, its numerous small fruits one-seeded, *alim* (*Melanolepis multiglandulosa*, fig. 82), *binónga* (*Macaranga tanarius*, fig. 72), *hinlaúmo* (*Mallotus ricinoides*, fig. 77), its fruits densely crowded and covered with soft processes, *banáto* (*Mallotus philippensis*, fig. 76), its small fruits covered with reddish scales, and *balánti* (*Homalanthus populneus*, fig. 74). These last five species are all shrubs or small trees belonging in the euphorbia family.

Soon the genera become considerably increased, including representatives of such groups as *balongái* and *batikúling* (*Litsea*) of the laurel family, *alagáo* (*Premna*) and even *moláve* (*Vitex*) of the verbena family, *aniláo* (*Colona*) of the linden family, various species of *Ficus* of the mulberry family, *anónang* (*Cordia dichotoma*) of the borage family, *takpó* (*Psychotria*) of the coffee family, *tanág* (*Kleinhovia hospita*, fig. 79), *bayóg* (*Pterospermum diversifolium*) with its large fruits which split open, bearing numerous winged seeds, *talósan* (*Helicteres hirsuta*), *baliknon* (*Melochia umbellata*), and *labáya* (*Commersonia bartramia*, fig. 78), the last four all belonging in the sterculia family. In places the common candlenut-tree *lúmbang* (*Aleurites moluccana*, fig. 81) is frequent and sometimes dominant. Its leaves are often pale green in contrast to the dark green of most tropical foliage.

The genus *Antidesma*, all shrubs or small trees, may be well represented by various species, one of the most common and widely distributed being *binayóyo* (*Antidesma ghæsembilla*, fig. 80), which is often characteristic of certain park-like open grass-lands. Worthy of mention here because of the poisonous qualities of its seeds, is the physic-nut (*Jatropha curcas*, fig. 84), although this introduced American species is more characteristic of hedgerows and the settled areas; its fairly large black seeds should never be eaten. Sometimes one finds certain of the wild yams, all vines, of the genus *Dioscorea*. Of these, *túngo* or *túgi* (*D. esculenta*, fig. 67) has slightly spiny stems, simple, broadly heart-shaped leaves, and tubers that are excellent food when boiled or roasted; forms of this are also cultivated. Another species known as *námi* or *karóti* (*D. hispida*, fig. 68) also has somewhat spiny stems, but its leaves are compound, each with three leaflets. Its tubers are poisonous and may best be left alone, although if they be sliced thinly, and the slices thoroughly soaked in running water and then cooked, the product may be eaten safely.

It is impracticable to consider all of the species that may be observed in these secondary forests, or for that matter, even all of the dominant ones. Recently established forests of this type are always very simple in that the number of different kinds of shrubs and trees is small. As they attain age, they become increasingly complex, and the total number of species of shrubs, vines, and trees greatly increases. As these secondary forests approach maturity, various species more characteristic of the primary forests may be observed, which have been developed by wind-, bird-, or animal-distributed seeds. Thus one may expect to find various representatives of the large genus *Canarium*, fig. 70, the mature seeds of all species being edible, and here and there the rather characteristic *bágo* (*Gnetum gnemon*, fig. 69), the fairly large seeds of which may be eaten when cooked, while its younger leaves are very widely utilized as a substitute for spinach; in fact, this tree has been called "New Guinea cabbage." It occurs all over the Malaysian region.

At the same time, especially in those regions where commercial contacts have been long continued, not infrequently one finds certain introduced and naturalized tropical American shrubs and trees dominating the scene, particularly the common guava (*Psidium guajava*, fig. 216) of the myrtle family, the seeds of which are widely distributed by birds and fruit-eating animals, and two representatives of the bean family (Leguminosæ), especially the *ipél* (*Leucæna glauca*, fig. 73), unarmed, with numerous small leaflets, fairly large globose heads of white flowers, and numerous thin brown pods; in places this tends to become dominant and gregarious. The other member of the bean family is the spiny *aróma* (*Acacia farnesiana*, fig. 75), with its small heads of fragrant yellow flowers and short thick pods. Often abundant will be the common lantana (*Lantana camara*, fig. 83), with its small but attractive varicolored flowers and somewhat aromatic leaves. This developed into a terrible pest in Hawaii, but in general cannot compete with the denser vegetation of the larger islands in the west. Epiphytes in these secondary forests are rare or may be entirely absent.

This is a temporary stage in reforestation, for gradually the small and medium-sized trees that make up these forests are overshadowed by the taller-growing species and are thus in time eliminated. A recently established secondary forest is on the whole a most disappointing place to explore from a botanical standpoint, for the simple reason that the bulk of the vegetation is composed of very few different species. As these forests attain greater age, they become more interesting because the number of different kinds of trees increases, but never do they present the intriguing interest that one always finds in the primary forests.

ENDEMISM IN THE SECONDARY FORESTS

One very striking contrast between the secondary and the primary forests, aside from the differences in the constituent genera and species and the relatively small trees in the former as compared with the latter, is the matter of endemism. An *endemic* species is one that is native of, but confined to, a definite geographic area, such as a single island or part of an island, while an *indigenous* species is one that has attained its present wider geographic range through natural means. The latter is, of course, a "native" species wherever it occurs but may be of a very wide geographic distribution. In both botany and zoology this distinction is important. Thus all of the widely distributed strand and mangrove forest species are indigenous no matter where they are found; many of them are actually of pantropic distribution, that is, occurring naturally in the tropics of both hemispheres, as distinguished from the man-distributed weeds and cultivated plants which owe their present distribution to the intervention of man as a disseminating agent. In the primary forests of Malaysia, as noted below, local specific endemism within any single island or any more or less compact island group, is very high and will average 75 per cent for all species that make up their total vegetation. In a recently established secondary forest endemism may be nil, but as these forests attain age the percentage will gradually increase, although nowhere will it be very high—certainly not over ten per cent.

We are forced to the inevitable conclusion that the primary forests in Malaysia represent the original vegetation of the region; that these great forests before the advent of man, or at least before agriculture became an established art, were much more extensive than they now are; and that a very high percentage of the constituents of the secondary forests represent those species that have greatly extended their ranges by natural means, and largely since the advent of man on the scene. Man has been a basic factor here, since through his destruction of the primary forests over vast areas the habitats were provided wherein those types of shrubs and small trees that make up the secondary forests not only established themselves but could reproduce themselves and thrive without being overshadowed and eliminated by the great trees that characterize the primary forests. Remove man from the scene, and in the course of some years the open agricultural lands would for the most part become secondary forests; and after a longer period the secondary forests would largely be replaced by the primary forests.

OPEN FORESTS IN DRIER REGIONS

While within those areas where the rainfall is abundant both the primary and the secondary forests are dense, in those regions where the rainfall is limited, more open types of both primary and secondary forests prevail. Thus in large areas in southern New Guinea, with a rainfall approximating that of eastern Australia, the trees are more widely placed, as in Australia, with open spaces covered with grasses and herbaceous plants, forming a more or less open and park-like terrain. Even within the wetter tropics, extensive areas will be noted here and there, covered with rather coarse grasses and with rather widely spaced, normally not very large trees giving such areas a distinctly park-like appearance. Here apparently, in spite of the heavier rain, the annually recurring grass fires inhibit the establishment of dense forests by the periodical destruction of seedlings and young plants of the arborescent species.

THE OPEN GRASS-LANDS

Throughout the region and in various types of topography, from open valleys to gentle or steep slopes, and at varying altitudes from sea-level to the tops of higher mountains, greater or less development of open grass-lands will be noted. These areas are normally dominated by few coarse grasses. In the alluvial valleys usually the wild sugar-cane, *gelágah* or *taláhib* (*Saccharum spontaneum*) forms dense stands six to fifteen feet high, its harsh leaves with cutting edges. On the slopes *álang-álang, lálang, kúnai* or *kógon* (*Imperata cylindrica*), mostly three or four feet high and much softer than the *gelágah*, occupies vast areas. Both of these coarse grasses have highly developed underground rootstock systems, which explains their persistence and dominance. In the more park-like savannah country the grass species will be found to be more numerous, representatives of such genera as *Panicum, Andropogon, Eragrostis, Themeda, Rottbœllia,* and others being present. At higher altitudes other coarse grasses will be noted, such as elephant-grass (*Thysanolœna maxima*), *Miscanthus,* and others, and on the highest mountains one not infrequently finds open heath-like formations, and even extensive development of grass-lands above the timber-line; but normally one has to ascend at least 10,000 feet, and in the deep tropics even higher, before reaching the upper limits of tree growth. At these higher altitudes most of the grasses and sedges will be representatives of widespread temperate-zone genera. In vast areas the forest cover is continuous from the seacoast to the summits of the mountains, particularly in sparsely populated regions.

ECOLOGICAL INTERESTS OF GRASS-LANDS

These open grass-lands, whether of natural origin or due to the influence of man in connection with his agricultural proclivities, are ecologically interesting, but from a purely botanical standpoint distinctly uninteresting, except those at higher altitudes, because of the very slight diversification of the flora. It is suspected that large areas now occupied by coarse grasses were originally forested, although undoubtedly here and there in Malaysia some natural grass-lands occurred before man disturbed the balance of nature as he developed his agriculture. There seems to be little doubt that many extensive grass tracts owe their presence, or at least their present extent, to the primitive type of agriculture long practiced by the native inhabitants. One reason for this belief is the very small percentage of local endemism, for wherever these grass-lands occur in the archipelago as a whole, the dominant species are the same throughout Malaysia; the seeds of the several most aggressive species are wind-distributed.

EFFECT OF RECURRING FIRES

Areas once occupied by the coarse grasses tend to remain as open grass-land for indefinite periods of time, chiefly because they are burned over each year, or every few years, in the dry season; and the very intense heat of these grass fires effectively destroys the seedlings and young plants of such trees as may tend to become established. Such open areas normally do not occur in regions where the rainfall is reasonably heavy in all or most months of the year, but are rather more characteristic of the regions having annual alternating prolonged wet and dry seasons. The more even distribution of the rainfall always favors the establishment and permanence of the forest as opposed to the open grass-lands; and yet open stretches of land once occupied by coarse grasses tend to remain as open grass-lands for indefinite periods of time.

PÁRANG VEGETATION

In the Philippines, where mixtures of open grass-land and second-growth forests occur, the descriptive word *párang* is used to define such areas; and this peculiar mixture, or alternation of grass-lands and second-growth forests, is just as characteristic of Malaysia as a whole as it is of the Philippines. These *párang* areas are usually found near the fairly densely populated regions. The forested areas may be relatively small, or at times rather strictly confined to the ravines and along watercourses; the coarse grasses tend to occupy the more gentle slopes. Scattered individuals of fair-sized trees may occur within the grasslands, apparently an indication that at least under some conditions the forest is attempting to reassert itself.

Fatal to this tendency to natural reforestation, even if such natural invasions by tree species are pretty well restricted to the types that characterize and dominate the secondary forests, rarely types characteristic of the primary forest, are, of course, the recurring and devastating grass fires. Here, as elsewhere in Malaysia, fires are usually intentionally set by the natives, although occasionally they may originate through natural agencies, by lightning or otherwise. It has been observed over and over again that where fires do not occur over a period of some years, the forest tends to reassert itself, clearly an indication that under natural conditions great areas of land now occupied almost exclusively by the coarse grasses would revert to forested areas. Man, with his purposely set grass fires, is the disturbing factor.

BAMBOO THICKETS

In some regions abandoned clearings may be occupied by dense thickets of thin-walled bamboos, these usually of the genus *Schizostachyum*. Such areas, after an interval, may be again cleared and utilized for agricultural purposes, even as are those that soon become occupied by the characteristic quick-growing small trees and shrubs above mentioned. Sometimes when large areas of primary forests are cleared or partly cleared in extensive lumbering operations, it may be bamboo that first dominates the scene, rather than small trees and shrubs. In the latter case this is apparently only the first phase of revegetation, for eventually the dense stands of bamboo will be replaced by the dominant trees that characterize the surrounding primary forests.

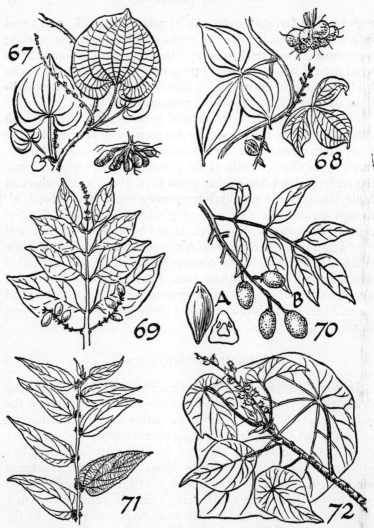

Secondary Forest Types: Fig. 67. *Dioscorea esculenta*; Fig. 68. *Dioscorea hispida*; Fig. 69. *Gnetum gnemon*; Fig. 70. *Canarium commune*; Fig. 71. *Trema orientalis*; Fig. 72. *Macaranga tanarius*.

SECONDARY FOREST TYPES: Fig. 73. *Leucaena glauca*; Fig. 74. *Homalanthus populneus*; Fig. 75. *Acacia farnesiana*; Fig. 76. *Mallotus phillippensis*; Fig. 77. *Mallotus ricinoides*; Fig. 78. *Commersonia bartramia*.

SECONDARY FOREST TYPES: Fig. 79. *Kleinhovia hospita*; Fig. 80. *Antidesma ghaesembilla*; Fig. 81. *Aleurites moluccana*; Fig. 82. *Melanolepis multiglandulosa*; Fig. 83. *Lantana camara*; Fig. 84. *Jatropha curcas*.

6

The Primary Forest

A primary or virgin forest is one where the original vegetation has remained through the centuries undisturbed by the activities of man. In the tropical insular region between Asia and Australia, virgin forests cover large areas. They are very complex in that the number of different kinds of trees is very great. They vary in their constituent species, depending on local climatic and soil conditions. Thus we have the brackish and salt-water forests along the lower reaches of tidal streams discussed in Chapter 4, those of the fresh-water swamps on the flood plains of great rivers, those of the rich well-drained soil of the great valleys, and those of the steep or gentle slopes. Local variations in the nature of the soil, exposure, drainage, the abundance and seasonal distribution of the rainfall, and other factors affect the picture. Another condition is the underlying geologic formation, for the vegetation of limestone areas is always very different from that found in volcanic regions. Thus a description of one tropical forest might be entirely different from what obtains in another area. Favorable conditions as to soil, moisture and other factors always result in an exceedingly complex forest dominated by very large trees. On sterile or very rocky slopes, even in regions of ample rainfall, the majority of the trees may be rather small. In all of these primary forests, most of the species are of very restricted geographic distribution, and from a strictly botanical standpoint are the areas most worthy of exploration.

EFFECT OF POPULATIONS ON FORESTED AREAS

The pressure of population has been a factor here, as in the relatively small island of Java, where, as the population has increased—and Java today, with a land area of about 48,500 square miles, corresponding roughly to that of New York State, supports a population of about 40,000,000 people—the demand for agricultural land has increased. The net result has been the destruction of the primary forest over most parts of Java below an altitude of about 4,000 feet. The clearing of great areas for the rubber, tea, cinchona, and other plantations for export commodities has also been a factor here, as in the Malay Peninsula and in other parts of the archipelago. It is thus in the less densely populated islands, and particularly the large ones like Sumatra, Borneo, and New Guinea, and smaller ones such as the Solomon Islands, Bismarck Archipelago, Celebes, parts of the Philippines, and others, that great stretches of primary forests still remain in an undisturbed condition.

CHARACTERS OF THE FOREST

When one enters the primary tropical forest from adjacent open country or even from adjoining tracts of secondary forest, the contrast is very great, often like stepping from one world into another. The scene is dominated by great trees with a mixture of a large number of smaller ones. Generally a three-story forest literally exists, particularly in regions subject to alternating dry and wet seasons, but these "stories" may not be so evident in constantly wet regions. The tall trees form the upper story, often with their great trunks unbranched to as high as 100 feet above the ground; these dominant trees form the canopy of the forest and represent many different species. The next story is composed of an even greater number of smaller tree species whose tops approximate in height the lower branches of the forest giants that form the canopy. Beneath this second-story series is yet another one composed of still smaller trees as well as shrubs, intermixed with seedlings and young plants of the larger trees, young rattan palms, and sometimes other representatives of the palm family, and in some islands various species of *Pandanus*.

In spite of the great diversity in species that make up these complex forests, they are not as a rule difficult to traverse. Great lianas may abound. One familiar with temperate-zone forests is usually impressed with the relative absence of fallen tree trunks, for in the humid tropical forests not only does dead timber decay rapidly, but the process of destruction is hastened by the presence of many different kinds of termites or white ants, which are very destructive to dead wood. The margins of primary forests, such as those along river-banks, may give the impression of impenetrability because of the lush overgrowth of lianas, but as a matter of fact, the actual forest interior is relatively open except for, here and there, tangles of climbing bamboos or rattan palms, or thickets of screw-pines (*Pandanus*). Physical characters of the terrain, especially the often deep stream channels that intersect the great flood plains and swamp forests, will make travel most difficult, but this does not apply to gentle or steep slopes.

GROUND-COVER

In those regions where annually a long dry season prevails, alternating with a long wet season, there is very little in the way of ground-cover such as herbs, grasses, and ferns; the soil is more or less covered with fallen leaves. In the more constantly wet forests, there is a distinctly greater development of herbaceous plants adapted to shade conditions, but grasses and sedges are usually absent. Normally, however, in both of these types of primary forest the canopy is so dense that one never sees the sun. Here and there one will catch a glint of sunlight on the leaves far overhead and thus will be able to determine whether the day is clear or cloudy.

In both types there is usually a certain number of lianas and climbing palms, especially in the genera *Calamus* and *Dæmonorops* (the rattan palms), and sometimes climbing bamboos, but it is generally only in ravines and along river-banks that colorful flowering vines will be observed. The climbing *Freycinetia*, a first cousin to the awkward screw-pine or *pándan* (*Pandanus*), may be represented by few to many species. It is usually conspicuous because of its red bracts and red fruits. In New Guinea its fruits form the favorite food of the bird of paradise.

There may be some epiphytic orchids, as well as some ground orchids, and both epiphytic and terrestrial ferns, at least in the more constantly wet areas; but these plants are usually rare in those forests where long dry seasons prevail, at least at low altitudes.

RELATIVE PAUCITY OF ANIMAL LIFE

There is often little in the way of bird and animal life in these great forests, and a solitary individual attempting to traverse them sometimes has a vague feeling of dread or fear inspired by the very extent of the forested areas and, at least in wet weather, the distinctly gloomy surroundings; and yet under anything approaching normal conditions there is little to fear and there should be little or no cause for alarm. After all, these great forests are, from many standpoints, distinctly pleasant, as the temperatures are lower than those which prevail in the open country. Where monkeys occur, they will usually make their presence known. Occasionally one will encounter wild hogs, and in some islands the native deer, but the latter prefer the secondary forests contiguous to open grass-lands. Among the birds, the raucous call of the hornbill, the monotonous cooing of pigeons, the shrieking parrots, and the call of the jungle fowl will be evident, while in New Guinea the harsh calls of the bird of paradise and of the cassowary will be heard. Such animal life as is present is naturally more evident along and near streams and near clearings. Very few snakes will be seen. Occasionally one may be vaguely disturbed by the sound of the fall of some forest giant tree, perhaps at a great distance, for such a sound carries in the forests. Here, as in other parts of the world, trees reach full maturity and eventually the end of their life span, and then fall with a disturbing crash.

THE PRIMARY FOREST AS AN ORGANISM

Pertinent to this general discussion are the following data from a recently published English translation of Dr. H. J. Lam's impressions regarding the primary forests of New Guinea on his trip up the Mamberamo River to the summit of Doorman Peak in the central mountain range in 1920.*

In its ideal form the tropical forest is an organism, the lower limit of which is the ground, the upper the forest roof. The latter is supported by what may be designated as the forest skeleton; the trunks of giant trees, frequently without branches, which rise several score of meters, their crowns forming the forest roof. But there is still a third limit, the margins, such as we find along the river banks. Both the roof and the margins are closed to such an extent that entirely different conditions prevail within the forest from those outside where the sun, the wind, and the rain have free play. Within the forest the light is dim, frequently suggesting twilight, the wind as well as the light is broken, and the rain loses its force. This is why, within the forest, all outward conditions remain within very strict limits. And while outside all sorts of extremes of light and dark, of wind and calm, of extreme heat and sudden cooling, of great humidity and dryness, are possible, yet within the forest these conditions are extremely constant, as if the inner parts of the forest were a natural conservatory wherein the light alone changes with day and night. For not only is the forest roof thick, but also even in still stronger measure, the forest-margin is closed by the overwhelming mass of lianas which frequently extend to the ground, and which can form so thick an overgrowth that—unfairly, as we shall see—the forest itself has been called impenetrable.

THE STRUGGLE FOR EXISTENCE AMONG PLANTS

In such a rain-forest with its conditions optimum in many respects, the struggle for existence among individuals and species is so sharp that, with an equal degree of justification, one can defend the proposition that these species and individuals, which carry on a life and death struggle, are units, while the primeval forest in its entirety may be looked upon as

* LAM, H. J. Fragmenta Papuana (Observations of a Naturalist in Netherlands New Guinea). Sargentia 5: 1-196, fig. 1-32. 1945. (Arnold Arboretum, Jamaica Plain, Mass.).

a mighty organism, wherein, as in all organisms, the maintenance of equilibrium is constantly sought. For if the balance becomes disturbed somewhere, for example, through the crash of an old forest giant, which in its fall has torn open a gap in the forest, and has broken the forest roof, and therefore has greatly changed the conditions in that spot, then we see immediately how the flora reacts to the infliction of the wound. . . . In the injured forest there immediately arises in such a place a dense mass of tall herbs, shrubs and young trees which now, for the first time, have an opportunity to develop. At the same time the loose flexible branches of lianas descend from the forest roof slowly to join with the lower flora and thus close anew the forest roof. . . . In this struggle a few plants slowly triumph, in the long run not more than one or two trees, and thereby the wound is healed. Meanwhile fungi and bacteria have done their work, completely destroying the fallen trunk.

MEMORIES

Many memories bind us to each place that we visited. The Mamberamo navigators will ever see before them the broad muddy river with its numerous curves, the still, high forest-walls, the translucent morning mists which hang without motion in the treetops until eight or nine A.M. They will feel the heat vibrate above the river banks at midday, and again sniff the heavy damp odor of the forest, evidence of the incessant progressive decay of organic material. Again the melancholy cooing of doves will be heard at regular intervals high up in the treetops, frequently also the noisy cry of the hornbills, which are hardly seen amongst the mass of leaves and branches, until with the harsh flapping of their wings, they fly away. Then we see again their dark silhouettes before us as they depart, frequently in pairs, over the river, as well as the screaming cockatoos with their blunt heads and swift wing-beat, which in the evening at the fall of twilight come in large flocks to sleep in the trees. Some will bear in memory the distant hoarse cry of the cassowary. Sometimes too it is the distant howl of a Papuan dog that comes hovering over the water by the river.

A TROPICAL STORM

Or his description of a cloudburst: A thick curtain of pouring rain came through the thatched roof, permitting nothing to remain dry. Here we sat hours on end, altogether motionless. Darkness fell early. Opposite us, from a terrace about 10 m. high, wild waterfalls thundered down. From near at hand we heard the landslips rattle, and at times in the neighboring forest an enormous tree crashed to the earth with a dull roar with an ominous cracking of the wood, which sounded above the noise of the waterfalls. With great difficulty a candle is lighted, and by its flickering light we see outside the naked shining bodies of the Dyaks busy pulling the boats higher and bringing the cargo to safety, fearful lest everything would drift away with the sudden rise of the water. Later the rain lessened a little and we tried to make a fire, but its thick smoke had no effect on the swarms of mosquitoes. . . . Thus the really small discomforts of the explorer's life alternate with many moments of happiness, for when, on the following morning, we departed upstream in a decreasing drizzle and watched the sun slowly appear, the distress of the previous night was forgotten.

TROPICAL *VERSUS* TEMPERATE-ZONE FORESTS

The more open broad-leaved forests of the temperate zones have little in common with the tropical prototype. In an ordinary mixed deciduous forest, such as those of the Eastern United States, the total number of different kinds of trees is small; but in these very complex forests of the Old World tropics there are not only hundreds of different trees species, but literally thousands, if one considers the vegetation over any extensive area. If in a temperate-zone forest one might find ten or a dozen different kinds of dominant trees on an acre, in these highly complex tropical forests he may expect to find as many as 100 different species or even more on a similar area. Very rarely will any forest be found that is dominated by a single tree species. The large trees forming the upper story or canopy of the forests are fairly well scattered, but the smaller ones forming the second story will stand rather close. Everywhere will be noted the curious strangling figs, all belonging in the single genus *Ficus,* the group that is characterized by the banyan (see p. 93).

BUTTRESSED TRUNKS

Another type that will be strange to the observer will be those great trees with their characteristic wide-spreading, relatively thin buttresses extending far from the trunk at the ground level and upward along the trunk for ten to fifteen feet, the bole otherwise being cylindric. These buttresses are developments that assist the tree in maintaining an erect position in the forest, acting as supports to the trunk.

PAUCITY OF CONSPICUOUS FLOWERS

Generally speaking, large and conspicuous flowers will be absent, although not entirely so. Sometimes the flowers and fruits are borne directly on the tree trunks themselves, or on specialized leafless branches springing from the trunks and larger branches. This phenomenon is known as cauliflory (see p. 94) and is particularly characteristic of various sylvan species of the wild figs (*Ficus*), but does not apply to the strangling figs mentioned above. The forest canopy is so dense, however, that it is difficult to detect trees in flower or in fruit, other than those characterized by the peculiar phenomenon known as cauliflory mentioned above, except as one notes portions of these on the ground where they have fallen from the higher branches. When fallen flowers or fruits are noted, it is frequently difficult to determine from which of the numerous neighboring trees they came.

SOME CHARACTERISTIC GENERA

Within the limits of a work of this type, and considering the exceeding complexity of these tropical forests, it is impracticable to enter into much detail regarding the constituent genera and species. Within the larger islands of the Western and Southwestern Pacific regions, there are many thousands of different tree species. In such a genus as *Syzygium* (*Eugenia*) of the myrtle family, there may be as many as 600 or 700 different species already known; and in such a characteristic genus as *Ficus* of the mulberry family, even more. Some of these species may be of fairly wide geographic distribution, and occasionally individual species will be found that have extended their range all over Malaysia from Sumatra to the Philippines, New Guinea, and the Solomon Islands. Generally, however, we encounter the problem of specific endemism in that in most genera that contain any considerable number of species, certain individual ones are characteristic of Sumatra, of Borneo, of Java, of the Philippines, of Celebes, of New Guinea, or of other geographical areas, or they may be strictly limited to certain parts of any one of the geographic entities above enumerated, not being found elsewhere. There are many genera in this category other than *Syzygium* and *Ficus* mentioned above, including *Canarium* of the Burseraceæ, *Aglaia* and *Dysoxylum* of the Meliaceæ, *Palaquium*, representing the gutta-percha family, *Diospyros* of the ebony family, *Terminalia*, representing the Combretaceæ, *Litsea, Cryptocarya, Cinnamomum,* and other genera of the laurel family, *Calophyllum* and *Garcinia,* with yellowish milky juice representing the Guttiferæ, *Knema, Horsfieldia,* and *Myristica,* representing the nutmeg family, *Polyalthia, Goniothalamus,* and other genera of the custard-apple family, *Sloanea* and *Elæocarpus,* representing the Elæocarpaceæ, and many others.

Sometimes one will note on the ground characteristic acorns that remind him of the same type with which he is familiar at home; these will be the fruits of the tropical species of oak

(*Quercus*), although these tropical oaks differ from those of the colder climates in having entire rather than lobed or toothed leaves. Most of these belong in a specific group that is recognized by many botanists as a distinct genus, *Lithocarpus*. The tropical counterparts of the northern chestnut (*Castanea*) will be representatives of an allied genus, *Castanopsis,* some of the species having spiny burs reminiscent of the chestnut. But representatives of such characteristic northern groups as the maple (*Acer*), the willow (*Salix*), walnut (*Juglans*) and its allied genera, poplar (*Populus*), basswood or linden (*Tilia*), buttonwood (*Platanus*), tulip-tree (*Liriodendron*), elm (*Ulmus*), beech (*Fagus*), birch (*Betula*), as well as the northern evergreen types such as pine, spruce, fir, hemlock, and their allies, will be conspicuous by their absence. Floristically the tropical forests are literally a new world to sojourners coming from temperate regions.

THE DIPTEROCARP FORESTS

Particularly in the Sunda Islands and in the Philippines, but to a lesser degree in Celebes, New Guinea, the Solomons, and the Bismarck Archipelago, the primary forests are often dominated by great trees characteristic of a single essentially Indo-Malaysian family, the Dipterocarpaceæ. The chief genera are *Dipterocarpus, Anisoptera, Shorea, Hopea,* and *Vatica*. These genera are all highly developed in the number of distinct species, especially in the Malay Peninsula, Sumatra, Borneo, and the Philippines, but are poorly represented in Celebes and New Guinea and its adjacent islands, for there only a few representatives of the last four genera occur. The primary forests dominated by the forest giants of this family are so characteristic that they are usually spoken of as dipterocarp forests. The trees are all large, the great boles unbranched frequently up to 100 feet, and they supply the most important local building timbers wherever they occur. Their fruits, usually small but sometimes, especially in *Dipterocarpus,* fairly large, are provided for the most part with from two to five wings that give the fruits a gyratory motion in falling, but which normally do not distribute the fruits over any considerable distances, since they are too heavy for wide wind dissemination. Normally reproduction by seeds is confined to forested areas, as the seeds do not germinate in the open country, or if they do, the young plants fail to become established.

EUCALYPTUS FORESTS

In parts of New Guinea, the Bismarck Archipelago, the Moluccas, and as far north as Mindanao in the Philippines, in favorable localities one will note giant *Eucalyptus* trees, usually characterized by their thin, flaky bark; their firm leaves are always distinctly aromatic. A number of different species of *Eucalyptus,* which is essentially an Australian group, occur in the islands closest to Australia, and in some of the drier parts of New Guinea actually become dominant. There is only one species that extends as far north as Mindanao, but this same species, *Eucalyptus deglupta,* also occurs in Celebes, the Moluccas, and new Guinea.

SPECIFIC ENDEMISM IN PRIMARY FORESTS

One very striking fact about the primary forests of Malaysia is the enormously high specific endemism. While in the main most of the genera characteristic of the Sunda Islands and the Philippines may extend to New Guinea and the Solomon Islands, for the most part the individual species of the different geographic areas are distinct. Thus an undisturbed forest in Borneo, the Philippines, Celebes, or New Guinea will be found to be composed very largely of species mostly confined to one or the other of these geographic areas, or even to parts of any one area. There may be a few widely distributed species in certain genera, but only a few, and these usually occur along or near the seashore. This very fact of highly developed local endemism makes a detailed description of the primary forest covering any wide area an exceedingly complicated and difficult task, for one has to consider a vast number of different species. Within each geographic area above mentioned it will be found that approximately 85 per cent of all woody species—trees, shrubs, vines, and palms, and for that matter most of the herbaceous species, including the epiphytic orchids—will be local endemics; the ferns are in general much more widely distributed. This specific endemism is, of course, one of the reasons why we have to deal with many thousands of different species of trees when we consider even in a most general way the primary forests of any part of the Indo-Malaysian region. The situation can in no way be compared with the simple forests that characterize the north temperate zone, for, as has been indicated elsewhere, in these tropical regions, outside of the mangrove forests, a gregarious type of forest—that is, one made up of a single species—does not exist, except in certain regions as in parts of New Guinea where large tracts of fresh-water swamp forests occur in the great river valleys dominated by such species as the sago palm (*Metroxylon*), *Melaleuca leucadendron* of the myrtle family with its pale papery bark, *Mitragyna speciosa* of the Rubiaceæ, and *Campnosperma macrophylla* of the Anacardiaceæ.

EVERGREEN *VERSUS* DECIDUOUS TREES

These great tropical forests are evergreen in that most of the constituent species do not shed their leaves at any one time as do the broad-leaved trees in the temperate zones at the approach of cold weather. It is true that old leaves fall, but this occurs throughout the year; in most species as senile leaves fall they are replaced by new ones, this gradual replacement occurring at all seasons. Leaf fall is much more marked seasonally in those regions where long dry seasons prevail, for this in turn is associated with the available water supply. In the tropics it is thus not temperature but moisture that determines whether an individual species be evergreen or deciduous or partly deciduous; for scattered here and there in the primary forests of the alternating wet and dry regions are individual species that are definitely deciduous in the dry season. One typical of this group is *malasápsap* (*Gyrocarpus americanus*), of the Hernandiaceæ, with its characteristic two-winged fruits, and another the giant *tagalinau* (*Gossampinus malabaricus*), of the Bombaceæ, its very large red flowers being produced after the leaves fall. And yet as this species occurs in the constantly wet areas it may be non-deciduous.

THE MID-MOUNTAIN FORESTS

As one ascends the forested slopes, a distinct change is observed in the general characters of the primary forest at altitudes varying from 1,800 to 2,000 feet above sea-level, or perhaps in some regions even higher. The large trees that dominate the rich forests at lower altitudes are replaced by different and usually smaller species. The characteristic tree ferns that may occur in some regions at low altitudes may become abundant, and their graceful fronds and peculiarly marked trunks add to the interest of the scene. These mid-mountain forests differ from the low-altitude ones, aside from the differences in individual species, in that they are essentially of a two-story type as contrasted with the three-story type that characterizes the low-altitude forest. Only occasionally will trees be noted that attain a height of more than 75 to 100 feet; the great majority of them are perhaps 50 feet high or even less.

In those geographic areas where species of the Dipterocarpaceæ occur, they disappear at these higher altitudes. While otherwise in general composition as to genera there may be no very striking differences aside from the elimination of the dipterocarps, the species in the same genus, be it *Aglaia,* or *Ficus,* or *Lithocarpus,* or *Syzygium,* or *Terminalia,* or any other group that is represented in these mid-mountain forests, are generally different from those which occur at lower altitudes in the three-storied forests. Representatives of certain groups found in the low-altitude forest become more numerous in individual species, and to a certain degree in numbers of specimens, while here and there will be observed representatives of other genera which scarcely or very rarely have representatives in the tall forests of the plains and lower slopes. Among the larger trees one is apt to note the great trunks of the dammar (*Agathis alba*), a tree with broad thick leaves and large cones, which is a tropical counterpart of the Kauri pine of New Zealand and the true pines of more northern regions. Yet in its essential components, this mid-mountain forest is still a tropical one, for in general the temperate-zone elements do not appear until even higher altitudes are attained.

Mid-mountain forests naturally vary, and a detailed description of the constituent elements of such a forest in one major island or group of islands might not apply to what may be observed in other places. Thus in parts of New Guinea the forest may be essentially made up of oak trees (*Quercus* or *Lithocarpus*) and the very closely allied *Castanopsis*, these overtopped by a scattered stand of very tall *Araucaria* trees; but *Araucaria*, which is a genus botanically allied to our pine trees, does not extend northward and eastward beyond New Guinea. Again, in New Guinea there may be a slight to plentiful mixture of larger trees representing genera more characteristic of lower levels. Here also the line of demarcation between the mid-mountain forest and the extremely mixed forest of the lower slopes is often very sharp, corresponding with the lower edge of the cloud banks which envelop the mountains almost daily. When entering such a mid-mountain forest from below, one often notices at once its two-story structure, its open character under the canopy, relatively sparse undergrowth mainly of ferns, slender shrubs, and treelets, and a pronounced scarcity of climbing plants, epiphytes, and mosses. Here again, as contrasted with the lowland primary forests, the tree trunks are not buttressed. In some regions, such as the Solomon Islands, the oaks are absent, and even in the Philippines, while some may be present, they are not dominant.

The line of demarcation between the low-altitude high forest and the medium-altitude mid-mountain forest may be sharp or rather gradual, for much depends on topography, presence or absence of ravines, steepness of slopes, and other factors, and yet the change is so great as to impress itself on even the casual visitor. In addition to the changes in the components of the forest cover or canopy itself, ferns and herbs may become more abundant particularly as one attains higher altitudes, the former especially showing great diversification in genera and in species. Epiphytic orchids and ferns may increase in number and in diversity as to genera and species, as well as the smaller vines and the shrubby undergrowth; still, as in other types of Malaysian forests, few grasses and sedges will be noted, for these are essentially plants adapted to growth in the open the tropical forest being too dense for them. For in these lower mid-mountain forests, even as in the taller low-altitude forest, the amount of light that finds its way through the dense canopy is still relatively small.

THE MOSSY FOREST

Above the mid-mountain forest and usually beginning at an altitude of about 3,000 feet or somewhat higher in some regions, the mossy forest commences, and this continues upward to the limits of arborescent vegetation. While evident even on the gentle slopes, this characteristic type becomes more and more impressive on the exposed ridges. Everywhere, on the ground, on ledges and boulders, on the trunks and branches of trees, the available space is occupied by a great profusion of mosses, scale mosses, lichens, ferns, orchids, and other small flowering plants. Individual specimens of almost any species represented in this complex may be exceedingly abundant. At all seasons, except occasionally when there may be a short dry spell, the thick blanket of moss is saturated with moisture. In most parts of Malaysia the exposed ridges and peaks of the medium-sized and higher mountains are regions of heavy precipitation; and even when rain does not prevail, the upper parts of the mountains are shrouded in moisture-laden clouds for days at a time or for certain parts of almost every day. The terrain has no chance of being thoroughly dried out, and hence this condition of continuous humidity favors the growth of the dominant species whether these be the lowly mosses or lichens, the numerous orchids and ferns, the often luxuriant herbaceous plants, or the vines, shrubs, and trees that make up this forest. In well-developed mossy forests a twig no thicker than a pencil may appear to be an inch or more in diameter merely because of the enormous development of the moss flora, and at times the smaller forms in this group will be noted as even extending over the surfaces of thick leathery leaves of the shrubs and trees.

Because of the constantly moist conditions, the epiphytic vegetation is very striking and dominant. This is made up of scores of genera of orchids and ferns, some small and delicate, others robust and coarse. The number of individual plants of both ferns and orchids is unbelievably great, and there is an enormous diversification in species in both groups. At the same time, terrestrial representatives of both groups occur; among the ferns especially, the sprawling masses of *Gleichenia* and the stiff erect stems of *Oleandra* form very dense thickets. Here also will be found the strange *Lecanopteris,* its very large, swollen, irregular basal portions being hollow and inhabited by myriads of small ants. For the most part, the orchids, although very abundant and frequently with very delicate flowers, are not of the type prized by orchid growers, who, in general, prefer those with large or extraordinarily large flowers; most of these mossy forest orchids have small or medium-sized flowers. A number of vines and shrubs, some with brilliantly colored flowers, will be seen; many of these belong in the Gesneriaceæ. In many places, climbing or scrambling bamboos, chiefly of the genera *Schizostachyum* and *Dinochloa,* form impenetrable or almost impenetrable thickets.

In these dank forests, constantly dripping with moisture, many strange plants will be found, including the curious pitcherplant (*Nepenthes*), the oriental counterpart of the North American *Sarracenia,* except that the latter is a low herb growing in bogs and the *Nepenthes* is a vine clambering through the dense thickets. Here the observer will note characteristic representatives of northern genera, including many showy species of *Rhododendron* and relatively inconspicuous species of *Vaccinium.* Then, too, raspberries (*Rubus*) may be represented by a number of different species, as well as violets, and other genera characteristic of the north temperate zone. Although we are still in the tropics geographically, the prevailing temperatures at and above altitudes of about 3,000 to 4,000 feet are more those of the warmer temperate regions than of the tropics. As

THE PRIMARY FOREST 87

one attains higher altitudes, this change in temperature becomes more and more noticeable, and at the same time representatives of more and more north-temperate-zone genera appear. At these higher altitudes, for the first time perhaps, in spite of the strange surroundings of the tropical mossy forest, the botanist familiar with northern plants commences to feel at home.

THE ELFIN-WOOD

On the exposed peaks and knife-edge ridges of the higher mountains, one will encounter the very dense elfin-wood—almost impenetrable thickets of small shrubs with wiry stems, often interlaced with vines of one type or another. The shrubs stand so close that it is almost impossible for one to force his way through the thickets without first clearing a path by vigorous use of a machete. Finally on the very high mountains one may emerge, at altitudes of from 8,000 to 10,000 feet or somewhat higher, in open heaths and grass-lands above the timber-line.

TEMPERATE-ZONE ELEMENTS

If one does reach such high altitudes, he will be even more impressed with the temperate-zone elements, for in addition to various grasses and sedges either identical with or reminiscent of those found in northern regions, he may encounter representatives of such groups as cinquefoil (*Potentilla*), buttercup (*Ranunculus*), *Veronica,* various northern genera of the Compositæ, gentians (*Gentiana*), and many other groups that are entirely lacking in the lower altitudes of the tropics. Here also, if he be at all familiar with the Australian flora, he will note a certain number of Australian types, for infiltrations have taken place at these high altitudes, both from the north (Asia) and from the south (Australia); and as very many representatives of the north temperate zone have reached as far to the south and east as the high mountains of New Guinea, so a smaller number of Australian types reached the high mountains of New Guinea,

Celebes, the Philippines, but to a definitely less degree those of Borneo, Java, and Sumatra.

REGIONS OF SPECIAL BOTANICAL INTEREST

To the individual who is interested in the flora of a single region, or the botanist or collector who is assembling material on which a later technical consideration of all or a part of the plants of a region may be based, it is the primary forest and the medium and higher altitudes that should claim his chief attention. It is from the vast primeval forests, no matter what the altitude may be, that most of the new species will come—not from the open grass-lands and the settled areas, and not from the secondary forests or from the vegetation immediately back of the seashore. Most of the species that will be found in the last four plant formations will prove to be of very wide geographic distribution, long since named and described from other regions and botanically well known; but primary forest collections from any previously unexplored area, dominated by the undisturbed original vegetation, will always yield many new, unnamed, and undescribed plants. Java as a whole, much of the Malay Peninsula, and large parts of the Philippines have been extensively explored and the resulting collections intensively studied. While much work has been done on the floras of Sumatra, Borneo, Celebes, the Moluccas in general, parts of New Guinea and the Solomon and Bismarck groups, vast areas in all of these regions have never been visited by a botanist or a collector; even today the Admiralty and New Hebrides groups are for the most part literally *terræ incognitæ* from a botanical standpoint.

FALSE FIRST IMPRESSIONS

At first sight, the casual visitor may be disappointed in the tropical forests, which will impress him as an unending mass of green foliage unrelieved, in general, by conspicuous and highly colored flowers *en masse*. He may reach the erroneous conclusion that these forests contain little of interest. As he commences to observe more closely, he will note that the flowers of a high percentage of the numerous species are small and relatively inconspicuous. In many cases they will scarcely be visible because of the very density of the forest canopy, for after all, as far as the trees are concerned, the flowers are mostly far overhead on the upper branches of the trees. In any preliminary conclusion that the observer may draw as to the apparent paucity of flowers he will be in error, for in no part of the entire world are there areas richer in genera and in species of flowering plants than in these great forests of Malaysia.

CYCLIC FLOWERING

To exhaust the possibilities of any tropical flora involves field work through all months of the year. In practice one may return to the same area year after year and still expect to find something new or hitherto overlooked, for curiously not all of the mature trees flower annually, but some produce flowers and fruits only at intervals of several or even many years. This cyclic flowering is little understood but it is a definite phenomenon. For example, while certain species of bamboo, particularly the climbing bamboos, may flower each year, others produce flowers only at long intervals; in fact, this phenomenon of intermittent or widely spaced bamboo flowering is so marked that many natives who have lived all their lives in a particular place will assure one that certain bamboos never produce flowers. In the case of one species, it is definitely known that the flowering period is in cycles of about twenty years. Therefore the botanist or collector who observes bamboos in bloom should always make a serious effort to secure specimens if possible, for only the vegetative parts of a considerable number of species are actually known. Foresters and botanists familiar with the characteristic dipterocarp forests of Malaysia will say on occasion, "This is a dipterocarp year," which means that, for some obscure reason, most or all of the numerous species in this family burst into flower. It is an observed fact that while some species produce flowers and fruits in abundance every year, others for some unknown reason vegetate year after year with no appearance of flowers; and yet when the favorable year arrives, all of the species will produce flowers and fruits in great abundance.

Those who are familiar with the vegetation of temperate regions take the seasonal changes for granted—a period of flowering in the spring and summer, the production of fruits, the harvest season, the leaf fall of deciduous trees, and finally the long cold winters when all annual plants die and perennial plants enter a dormant or resting period. This is accepted as normal because it is a series of phenomena thoroughly familiar to us. When, however, we enter the tropics, the scene changes. In the humid, as opposed to the arid or semi-arid tropics, there may be differences in seasons, such as a wet season regularly followed by a dry season, but in other regions the rainfall may be fairly well distributed through all months of the year. This seasonal distribution of the rainfall is the important factor that governs the seasons, rather than temperature, and at the same time strongly affects the general types of vegetation.

In the tropics, except in areas of scant rainfall, growth is perpetual, and there is no real resting season, except for the partial one characteristic of regions that regularly have protracted dry seasons. As one result, everywhere in the tropics some species will be found in flower in one season, some in another, while other species may be found in bloom in all months of the year. This matter of flower production is clearly associated with such seasonal changes as may obtain here and there in the tropics. Yet without regard to the problem of seasonal distribution of the rainfall, it is worthy of note that the majority of the species that characterize the mangrove forests and the strand vegetation flower continuously.

REDUNDANCE OF VEGETATION

Most of the trees and shrubs are evergreen, and whenever one views the primary forests, and even the secondary ones for that matter, from the outside or from within the forest itself, he is impressed with the redundance of vegetation, the great variety of plants, and the paucity of showy flowers. In ultimate analysis, however, this massed and continuous tropical vegetation is the generalized type, as opposed to the specialized type of temperate regions; for in the latter, because of radical changes in temperature as between the summer growing season and the winter resting stage, most of the plant species have adapted themselves to these very radically different seasons through developing the phenomenon of deciduousness or leaf fall. Thus the vegetation of the colder parts of the earth has become specialized to meet the rigors of the winter season. Even when plants are evergreen in the colder temperate regions, such as in the Coniferæ or pine family, which includes the pines, spruces, firs, hemlocks, cedars, junipers, and allied groups, adaptation to unfavorable climatic conditions has taken the form of stiff narrow leaves which are hard in texture; but deciduousness extends even to this family in the larches (*Larix*) and in the allied genus *Pseudolarix*. Deciduousness of broad-leaved trees in the tropics is a response not to temperature but to deficiency in moisture, and in regions where long dry seasons prevail some of the trees are leafless for shorter or longer periods of time; but the percentage of deciduous trees is so small that this phenomenon scarcely affects the massed green impression that the tropical forests always present.

DIFFERENCES IN TROPICAL AND TEMPERATE-ZONE SPECIES

It is a patent fact that the tropical flora is made up almost entirely of species utterly different from those that thrive in temperate regions, although here and there one may note a weed or a cultivated plant that seems to be just as much at home in the tropics as in temperate zones. While a very few species of palms will thrive in certain warmer parts of the temperate zones, the group is enormously developed in tropical regions. No matter what their size, be they pygmies like certain species of *Iguanura* and *Pinanga,* or giants like the *talipot* or *búri* (*Corypha*), or enormously long vines like the rattan palms (*Calamus* and *Dæmonorops*), even the tyro will always recognize them as palms, regardless of the enormous variation in vegetative, floral, and fruit characters, for the palms form one of the most striking of the natural families of plants. Again, in the tropics, the bamboos are always impressive, although, like the palms, they vary greatly in size and in habit, ranging from small plants a foot or two high to giant grasses perhaps 100 feet tall, with various climbing species.

Everywhere in the tropics, as in the less-favored temperate regions, the vegetation responds to the environmental conditions. Literally in the vast primary forests of Malaysia one gains the impression of being in a great natural conservatory or hothouse, because of the temperature, the humidity, some of the types of plants that grow naturally in such regions, and even the odors inherent in the humid tropical forest. The differences, as compared with a conservatory, will be that the majority of the plants are huge trees and that within these forests the light will be relatively dim and the total number of herbaceous plants relatively few.

7

Noteworthy Plants of Special Interest

STRANGE ADAPTATIONS

Almost immediately the observer will note strange adaptations, such as in the giant lianas clambering here and there, some with peculiarly flattened stems, the rattan palms with their smooth stems of the same diameter throughout their great length of sometimes several hundred feet, and their striking adaptations for climbing through the long whip-like flagellæ armed with stout sharp recurved claws which are either an extension of the midribs of the leaves or are produced from the inflorescences. It is through this specialized adaptation for climbing that these strange non-twining palms ultimately succeed in attaining the height of the tallest trees, for when one is so placed that he can scan the forest roof, he will note the crowns of these palms extending here and there above the tree tops. On the trunks of many of the trees he may note peculiar smaller vines that grow closely appressed to the tree trunks and have juvenile forms with leaves utterly different from those of the mature ones, as in such groups as *Pothos* of the calla lily family, *Adenia* of the Passifloraceæ, and the climbing fern *Teratophyllum*.

THE STRANGLING FIGS

Everywhere in these forests will be noted the strangling figs, various species of the protean genus *Ficus,* with literally hundreds of different species in the tropics of both hemispheres, but particularly abundant in Indo-Malaysia. Here the plantlet starts from a small seed casually dropped by some bird or animal high on the branches of some forest giant. At first it is a harmless little epiphyte. Soon it commences to send its branches upward and its closely appressed roots downward, and from the time growth commences, the giant tree on which it began its development is doomed. Gradually the roots extend downward until they reach the earth, fuse, and eventually strangle the host plant, after which the latter disintegrates. In its place will stand a giant fig tree often characterized by its irregular trunk, made up of the fused or partly fused roots of this strangler, and sometimes supported by wide buttresses. It is not surprising that the natives of Malaysia have various superstitions regarding these strange and utterly valueless trees.

CAULIFLORY

Another phenomenon that will attract attention is cauliflory. This is the production of flowers not on the slender twigs of the ultimate branchlets far overhead, but in quantity on the trunks and larger branches of the trees themselves. Osbeck, who on a voyage from Sweden to China noticed this phenomenon during a short stop in Eastern Java on January 20, 1752, was so surprised that he actually described what he observed as a leafless parasitic plant under the name *Melia parasitica*, naïvely commenting: "A small herb of barely a finger's length growing on the tree trunks. It is so rare that, so far as is known, no one ever saw it before." Cauliflory is, of course, a phenomenon that never occurs in northern Europe, and Osbeck's assumption that what he observed—a short leafless inflorescence produced from the bark of the tree trunks—was a parasite, was but natural. What he saw and described was the cauliflorous inflorescences of the cultivated fruit tree known as *dúku, lángsat,* or *lansóne* (*Lansium domesticum,* fig. 204). Other cultivated fruit trees bearing their flowers and fruits directly on the trunks and larger branches are cacao or chocolate (*Theobroma cacao,* fig. 209), *kembóla* (*Averrhoa carambola,* fig. 203B), its smooth very acid fruits reminding one of small cucumbers, the *nam-nam* (*Cynometra cauliflora,* fig. 87), a small tree belonging in the bean family with strikingly paired inequilateral leaflets and edible acid fruits, and above all the spectacular representatives of the breadfruit genus *Artocarpus,* with their massive fruits sometimes weighing as much as fifty pounds, the *nángka* or jakfruit (*Artocarpus heterophylla,* fig. 85), and the allied *chempédak* (*A. champeden*).

It is, however, particularly within the primary forests that one notes the most species characterized by this phenomenon of

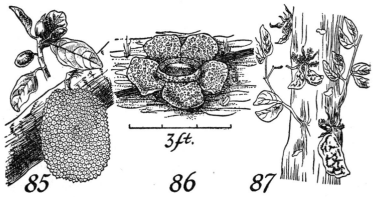

CAULIFLORY: Fig. 85. *Artocarpus heterophylla*; Fig. 86. *Rafflesia arnoldii*; Fig. 87. *Cynometra cauliflora.*

cauliflory. These include numerous species of wild figs (*Ficus*) —not, however, of the strangling type mentioned above— various species of *Syzygium* (*Eugenia*) of the myrtle family (Myrtaceæ), some species of *Dysoxylum* of the Meliaceæ, occasionally a species of *Diospyros* of the ebony family (Ebenaceæ), perhaps a really spectacular representative of the Dilleniaceæ (*Dillenia cauliflora*) with its large yellow flowers actually lighting up the gloomy forest, various representatives of *Phaleria* of the Thymelæaceæ with its slender pink or white flowers, occasionally a species of *Canarium* of the Burseraceæ, or even such a spectacular tree as *Semecarpus gigantifolius* of the Anacardiaceæ, its slender, tall, unbranched trunks bearing its flowers and fruits throughout the length of the trunk, which bears at the summit a crown of very long simple leaves reaching a length of five or six feet. Various genera of the custard-apple family (Annonaceæ), such as *Goniothalamus* and *Polyalthia*, present this same phenomenon, and have often fairly large flowers.

It is, however, within the genus *Ficus* that one notes the greatest variety in this strange adaptation, for in some species the clusters of fruits are borne on tubercles scattered here and there on the trunk; these tubercles are sometimes at the very base of the tree. Sometimes the fruits are produced on variously branched but always leafless inflorescences, which are often pendulous and simple or divided; they may be up to fifteen feet in length, hanging from the upper parts of the trunk or from its larger branches. In still others, the fruits may be borne on specialized branches radiating from the base of the tree and partly covered with earth and dead leaves. In one species (*Ficus mirabilis*), these leafless basal fruit-bearing parts are simple and up to ten feet long. Some of the smaller shrubs that grow in the forests also produce their flowers and fruits on their trunks. This phenomenon of cauliflory is an adaptation to environmental conditions and occurs in a considerable number of totally unrelated natural families of plants. It is so striking and so unexpected that it always attracts attention.

SYMBIOSIS BETWEEN PLANTS AND ANTS

Another phenomenon of unusual interest is that of symbiosis between plants and ants. This association is of mutual advantage to both the plants and the insects, for the plants derive a certain amount of nourishment from the detritus deposited by the ants and the latter find a home within the living plant itself. The word "symbiosis" merely means the living together of two dissimilar organisms in more or less intimate association. Among the most striking of the adaptations are the strange epiphytic species of *Myrmecodia* (*M. tuberosa*, fig. 96) and *Hydnophytum* of the coffee family (Rubiaceæ), some of which occur at low altitudes and sometimes even in the mangrove forests, while others grow in the primary forests at high altitudes. The basal parts of these plants, sometimes armed with short spines, are very greatly enlarged, and this enlarged part is tunneled by a series of large passages, from some of which small openings to the exterior occur; within the greatly swollen bases of these plants myriads of small black ants find their abode. Springing from the top of the tuber-like, tunneled base are the vegetative parts of the plant, sometimes stout and unbranched, sometimes rather slender and much branched; the small white flowers and the small fleshy fruits are borne in the leaf axils or in the axils of fallen leaves. In this same family (Rubiaceæ) one notes,

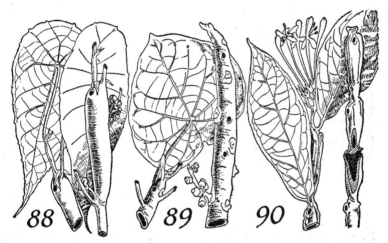

SYMBIOSIS WITH ANTS: Fig. 88. *Macaranga caladifolia*; Fig. 89. *Endospermum formicarum*; Fig. 90. *Clerodendron fistulosum*. All after *Beccari*.

especially along swift streams, representatives of such genera as *Neonauclea* and *Myrmeconauclea,* where the internodes of slender branches are perforated, more or less swollen, and hollow, with each hollow internode occupied by small ants. Similar adaptations are noted in other families such as in *Endospermum moluccanum, E. formicarum,* fig. 89, and *Macaranga caladifolia,* fig. 88 (Euphorbiaceæ), *Clerodendron fistulosum,* fig. 90 (Verbenaceæ), a nutmeg, *Myristica myrmecophila* (Myristicaceæ), sometimes the swollen and perforated leaf sheaths of climbing palms (*Korthalsia*), and occasionally leaf adaptations to insect abodes in such genera as *Piper* (*P. myrmecophilum*) of the Piperaceæ and *Saurauia* of the Dilleniaceæ.

Perhaps the most peculiar of all, however, are those leaf adaptations noted in such groups as *Hoya, Dischidia,* and *Conchophyllum,* all vines with abundant milky sap belonging in the milkweed family (Asclepiadaceæ). Some of these hang free, as epiphytes or semi-epiphytes, but in *Conchophyllum* and in some species of *Hoya* the vines may be closely appressed to the tree trunks or branches, while the circular leaves, one row on

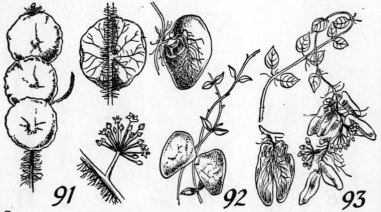

SYMBIOSIS WITH ANTS: Fig. 91. *Hoya imbricata*; Fig. 92. *Dischidia vidalii*; Fig. 93. *Dischidia rafflesiana.*

each side of the slender stem, are convex with their margins very closely appressed to the bark (*Hoya imbricata,* fig. 91). Under each leaf many roots are produced from the leaf axil which often quite cover that part of the bark protected by the leaf, serving to hold the plant in place and to absorb moisture and nourishment for the needs of the plant; each of these ready-made homes under each leaf is occupied by colonies of small ants. In other species of *Dischidia* and *Hoya* there are two types of leaves: the normal unaltered small ones, and those greatly specialized in that they are much swollen and hollow (*Dischidia vidalii,* fig. 92, and *D. rafflesiana,* fig. 93). Into these hollow leaves the axillary plant roots extend to absorb moisture and

NOTEWORTHY PLANTS

nutriment, for again these hollow leaves are usually the abodes of ants.

This phenomenon of symbiosis between plants and ants is by no means confined to the flowering plants, for it is also developed in certain ferns, notably in the case of the hollow creeping stems of the epiphytic *Polypodium sinuosum,* fig. 94, a species not uncommon on trees along tidal streams. At higher altitudes at and above 3,000 feet the strange *Lecanopteris carnosa,* fig. 95,

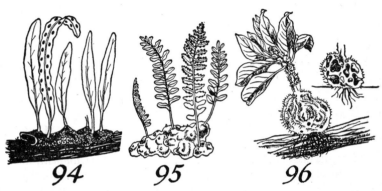

SYMBIOSIS WITH ANTS: Fig. 94. *Polypodium sinuosum*; Fig. 95. *Lecanopteris carnosa*; Fig. 96. *Myrmecodia tuberosa.*

is often conspicuous, with the greatly enlarged hollow irregular basal parts a foot or more in diameter. Both of these ferns provide homes within their stems for colonies of small black ants.

STENOPHYLLY

This term means "narrow leaves." In those regions where swift-running mountain streams occur that are subject to sudden floods in times of heavy rains, one may note a peculiar adaptation in that various shrubs, growing on the banks subject to sudden inundation for limited periods, are characterized by unusually narrow leaves. It is suspected that this phenomenon is associated with the habitat, and that the narrow leaves merely represent a special adaptation of the plants to a local condition.

The classical example is the *lumanái* (*Homonoia riparia*), a widely distributed shrub that grows in and along the beds of small, swift-running streams, but the phenomenon extends to a very considerable number of species in totally unrelated families, including *Atalantia* (Rutaceæ), *Syzygium* (Myrtaceæ), *Fagræa* (Loganiaceæ), *Ficus* (Moraceæ), *Garcinia* (Guttiferæ), *Psychotria* and *Neonauclea* (Rubiaceæ), *Saurauia* (Actinidiaceæ), *Tetranthera* (Dilleniaceæ), *Erycibe* (Convolvulaceæ),

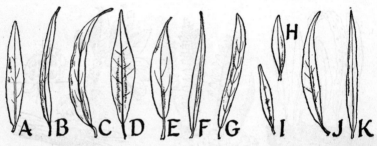

STENOPHYLLOUS PLANTS: Fig. 97. A. *Neonauclea angustifolia* (*Rubiaceae*); B. *Fagraea stenophylla* (*Loganiaceae*); C. *Garcinia linearis* (*Guttiferae*); D. *Syzygium neriifolium* (*Myrtaceae*); E. *Saurauia angustifolia* (*Actinidiaceae*); F. *Erycibe longifolia* (*Convolvulaceae*); G. *Homonoia riparia* (*Euphorbiaceae*); H. *Eugenia mimica* (*Myrtaceae*); I. *Atalantia linearis* (*Rutaceae*); J. *Psychotria acuminata* (*Rubiaceae*); K. *Tetranthera salicifolia* (*Lauraceae*).

fig. 97, A–K, and representatives of other families. In outline, texture, and other characters, the leaves of these stenophyllous species in botanically quite unrelated genera are strangely similar; so similar, in some cases, that at times one almost needs flowers or fruits in order to make safe generic determinations. Narrow leaves are, of course, by no means limited to species growing only in the indicated habitat. In totally unrelated genera the adaptation of these stream-bank species through the development of narrow leaves, in contrast to other species of each group characterized by broad leaves, is distinctly striking in this special habitat, and is of some biological interest as indicating how plant species may adapt themselves to very special conditions.

Flash floods, where suddenly and for brief intervals small swift mountain streams may become much swollen, are characteristic of the mountainous wet tropics. Under such conditions broad leaves would largely be destroyed to the detriment of the plant, but narrow hard leaves will withstand the strong current and submersion for some time. These sudden floods may vary from a few feet above the normal level of the stream to many feet; sudden rises of as much as twenty feet or more have been recorded within a half-hour period. When the heavy rains cease, the water level of the stream may fall as rapidly as it arose at the beginning of the torrential rain.

STRANGE PARASITES

If the individual be fortunate, he may see the giant flowers of the strange *Rafflesia*. A dozen species of this genus occur in Malaysia; some have flowers only ten to twelve inches in diameter, others bear gigantic ones—the largest individual flowers known being up to three feet in diameter. These large-flowered species (*Rafflesia arnoldi,* fig. 86, and *R. schadenbergiana*) occur in Sumatra and in the Philippines. This striking genus was named in honor of Sir Stamford Raffles, founder of Singapore. The plants are parasitic on the prostrate stems of relatives of the common grape vine, and occur only in the primary forests; they are very local. There are no stems or leaves, and the great round buds suggest cabbages in size and form but not in color. When the flowers open, there is no mistaking them, for the overpowering odor is like that of decaying meat. The colors are in general purplish or brownish purple, often more or less mottled.

Smaller brownish to purplish root parasites more frequently observed and also with no green coloring matter and quite leafless, are representatives of *Balanophora* and allied genera, and belong in a family allied to the Rafflesiaceæ. These plants grow in irregular clusters near the bases of tree trunks and, like the species of *Rafflesia,* are dependent on their host plants for food and water. They are much smaller than the species of *Rafflesia,* the individual flowers being small and densely crowded. Among the shrubby semi-parasites will be numerous species of the mistletoe family (Loranthaceæ), with greenish to yellow and red or purple flowers. Some of the species are rather conspicuous because of their relatively long flowers.

SAPROPHYTES

In the humid tropical forests various saprophytes will be observed. These are flowering plants with no green coloring matter, which depend for their nourishment entirely on decaying vegetable matter, quite like the even more numerous fungi. These plants vary in color from white to brown or yellowish brown, and various natural families are represented. Thus all of the slender species of the family Triuridaceæ, delicate plants with the color of dead leaves and hence difficult to detect, may be observed, a few representatives of the gentian family with white or blue flowers in rather striking contrast to their pallid stems, and even more numerous saprophytic orchids, the giant among these being a coarse rambling vine of the genus *Galeola*. The Polygalaceæ is represented by small nearly white plants of *Epirixanthes*, while some species of *Burmannia*, a group allied to the Orchidaceæ, may be noted.

SOME SPECTACULAR VINES

Among the vines one may see various representatives of the striking genus *Bauhinia* of the bean family, with peculiar two-lobed leaves, and masses of small to large, very irregular conspicuous flowers varying in color from yellow to red, or equally strange representatives of the genus *Mucuna*, their pendent racemes of rather small to distinctly large bean-like flowers ranging in color from greenish to yellowish red, bright red, or black-purple. In New Guinea and in various neighboring islands the large red flowers of the D'Albertis creeper (*Mucuna albertisii* and allied species) are particularly conspicuous along forested river-banks. In many of the species, fig. 2, the pods, often strikingly sculptured, are covered with very stiff, needle-like, irritating stinging hairs. In some regions the jade-vine (*Strongylodon*), of the same family, may be observed; the elongated hanging many-flowered racemes consist of numerous large jade-green to verdigris-colored flowers. Other species have bright red or blue flowers.

Among the vines with inconspicuous blooms, also in this same family, is the *gógo* or soap-bark-vine (*Entada phaseoloides*), the crushed stems of which are rich in saponin and are widely used as a substitute for soap. The striking thing about this vine is its enormous hanging, somewhat jointed pods, each section of which contains a single large, hard, shining chestnut-brown to reddish brown seed two inches in diameter. The pods may be up to three feet long and three to four inches wide. The seeds are the so-called sea beans, for the plant is of world-wide distribution in the tropics. The large seeds are often picked up on the seashores, for that matter, even as far north as Scandinavia, where they are transported by the Gulf Stream. Generally speaking, the vines with showy flowers will be observed chiefly along the margins of forests and along forested streams, for to produce flowers they need exposure to strong light, which is never present within the forest itself.

STRIKING SPECIES OF THE CALLA LILY FAMILY

Worthy of mention here are representatives of the strange genus of the calla lily family, *Amorphophallus,* of which some eighty different species are known. The "flowers" (really inflorescences composed of a spathe corresponding to the white part of the calla lily, and the spadix, corresponding to the inner yellow part, which bear very numerous small flowers on their basal parts) vary in size from a few inches in diameter to as much as four feet, and the individual inflorescences (spathe and spadix) of the gigantic Sumatran *krúbi* (*Amorphophallus titanum,* fig. 99) may reach a length of eight feet. The commonest and most widely distributed species is *badú, kébing bangáh,* or *pungápung* (*Amorphophallus campanulatus,* fig. 100), whose "flowers" are about one and one-half feet wide. In all of these species the "flower" (inflorescence) appears first, developed from the subterranean corm, and these have the same disagreeable odor as the species of *Rafflesia.* Following the appearance of the "flower," the vegetative parts develop—a tall, cylindric, usually rough and mottled stem and leaves divided into three parts which are again variously subdivided. Other representatives of this family are coarse vines which ascend along tree trunks within the forest, belonging to the genera *Raphidophora* and *Epipremnum,* sometimes known as *tibátib.* Here and there in rich alluvial soil one will note gigantic specimens of the so-called elephant's-ear or *badiáng* (*Alocasia macrorrhiza,* fig. 98), sometimes with dis-

UNUSUAL PLANTS: Fig. 98. *Alocasia macrorrhiza*; Fig. 99. *Amorphophallus titanum*; Fig. 100. *Amorphophallus campanulatus.*

tinct cylindric trunks as tall as a man, and surmounted by its immense leaves, and in fresh-water swamps the gigantic *paláuan* (*Cyrtosperma merkusii*), its very stout, greatly elongated, somewhat spiny leafstalks bearing the large leaves; the stalk and the leaf attain a length of twelve to fifteen feet. These, like *Amorphophallus titanum,* fig. 99, are the giants of the calla lily family.

GUTTATION

In the constantly humid primary forests where at all times, or at least through the rainy season, the humidity is very high, there is the strange phenomenon of the internal moisture of the leaf actually dripping from small pores at the tips of the leaves, particularly in representatives of the calla lily family. A high percentage of tropical species have leaves with long slender sharp drip tips, but this is normally only an adaptation for the prompt removal of superabundant surface moisture in the very wet weather. Guttation, however, is a special adaptation worthy of note. In the constantly humid forests there is little or no transpiration of moisture from the internal parts of the leaves. Without some special adaptation for the removal of this surplus moisture, the plant could not possibly thrive. This is attained by the phenomenon of guttation. This dripping of the internal moisture of the leaf compensates for the absence of transpiration. This insures a constant upward movement of plant sap which, of course, carries the nutrients on which plant life and plant growth depends.

PLANTS, INSECTS, BIRDS, AND ANIMALS

Anyone interested in the study of insects and birds knows by experience and observation that various plant species, when in full flower, attract many different kinds of these representatives of the animal kingdom. Thus a tree in full flower may form an important locus for his special purposes. The birds are in part attracted by the insects which they utilize as food, and in part by the presence of sweet secretions in the flowers; the insects are attracted by the pollen, by the sweet secretions, by colors and odors. All are familiar with the pleasing odors of many flowers, and some, of course, realize that disgusting odors, such as those of flowers of the *kalúmpang* (*Sterculia fœtida,* fig. 34), *Amorphophallus,* fig. 99, 100, *Rafflesia,* fig. 86, and representatives of other genera, serve their special purposes by attracting carrion flies. Some flowers are fragrant only at night, such as the *dama de noche* (*Cestrum nocturnum*), and some only or chiefly in the daytime.

Biologically, these visitors are most important to the life of the plant, for without pollination fertilization would not be effected, and hence there would be no seed production. There are very many curious adaptations in flowers to attract these floral visitors, and while much work has been done especially in the temperate regions, relatively few observations have been made in the tropics. There is thus a very wide field of observation open to the inquiring mind. These numerous interrelations of plants and animals are, of course, vital to both, for in general there is a very close correlation between plant and animal life in all parts of the globe where plants and animals thrive. Less obvious is the observed fact that in the tropics in those vines and trees where the large flowers hang in long stalks below the foliage and the branches, and in those trees where the inflorescences project well above the mass of leaves, pollination in some cases apparently depends on bats as visitors, rather than birds or insects. It may be that the bats are attracted by the colors or odors of the flowers, or they may, in some cases at least, be attracted by the insect visitors to the flowers.

NOTEWORTHY FERNS

The fern flora is so enormously developed that even a most simple treatment of the species alone would make an oversized volume. There are at least 1,500 different species known from the Malaysian region as a whole. They are particularly abundant in damp ravines at low altitudes and in the characteristic mountain forests at and above altitudes of about 3,000 feet. They vary in size from a fraction of an inch, including various representatives of the exceedingly delicate filmy ferns (*Trichomanes* and *Hymenophyllum*), to such tufted giants as *Angiopteris,* whose great fronds are sometimes ten feet long or more, and the numerous tree ferns of the genus *Cyathea,* whose trunks vary in height from a few feet in the smaller species to as much as forty feet in the taller ones. These trunks usually are strikingly marked by the large scars of fallen fronds. Many ferns are terrestrial; others are epiphytic. Most of them are erect, but others are vines. In a considerable number there is a striking dimorphism between the sterile and the fertile fronds; that is, the fertile fronds are entirely different in shape and in size from the sterile ones.

Among the epiphytic forms, the peculiar bird's-nest-fern (*Asplenium nidus*), with its great smooth entire leaves, is fre

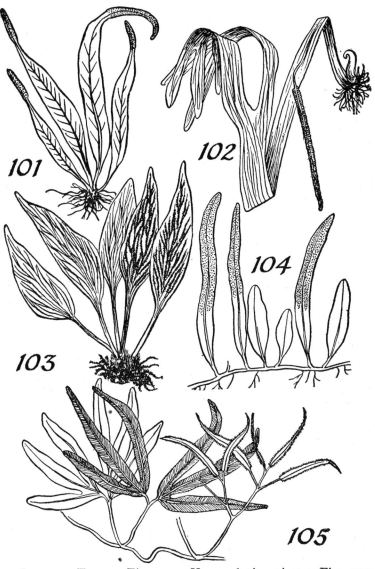

STRANGE FERNS: Fig. 101. *Hymenolepis spicata*; Fig. 102. *Ophioglossum pendulum*; Fig. 103. *Antrophyum plantagineum*; Fig. 104. *Cyclophorus lanceolatus*; Fig. 105. *Lygodium circinnatum*.

quently common. Forms allied to this species have leaves up to six or seven feet long. Sometimes one notes the really gigantic epiphytic species known as *Polypodium heracleum,* with its coarsely pinnate fronds, and again the even more striking staghorn-fern (*Platycerium*). The latter bears three different kinds of fronds—those forming the great cup-like structure that gathers and holds humus, the pendulous, forking sterile ones whence the name "staghorn," and the circular or lobed fertile fronds with their backs quite covered by the spore cases. The commonest of all of these epiphytes is the oak-leaf-fern or *pakpak láuin* (*Drynaria quercifolia,* fig. 110), so called because the more or less overlapping, persistent, at first green but eventually brown sterile leaves, the objective of which is to gather and retain humus, somewhat resemble large oak leaves. These persist as stiff brown bract-like organs long after they have ceased to function as leaves. The large, pinnately lobed, fertile fronds are entirely different from these sterile humus gatherers. In places, and sometimes associated with other epiphytic ferns, one will note in abundance the long, thin, green, pendulous, ribbon-like fronds of the epiphytic *Ophioglossum pendulum,* fig. 102, a strange "adder's-tongue" fern. The other species are terrestrial, erect, and much smaller.

Almost certain to be noted will be the *lagólo,* or *pakó-laut* (*pakó* = fern, *laut* = ocean) (*Acrostichum aureum,* fig. 66A), because of its occurrence near and in brackish marshes along the seashore—a habitat we scarcely associate with ferns—with its great tufts of leathery pinnate leaves, often as tall as a man. Also within the influence of salt and brackish marshes will be noted the climbing *diliman* or *hagnáya* (*Stenochlæna palustris,* fig. 66B); its tough stems are prized for binding purposes in structures such as fish traps, which must withstand the action of salt water. One will observe in thickets the often abundant climbing ferns of the genus *Lygodium* (*L. circinnatum,* fig. 105) commonly known as *nito,* and in the same habitats the strange relative of the adder's-tongue-fern, *ákan pakú, pakís kalér* (*Helminthostachys zeylanica,* fig. 106), with its erect stems, divided leaves, and central erect fruiting spike. A Philippine name used for this peculiar fern is *tonkúd lánguit,* which

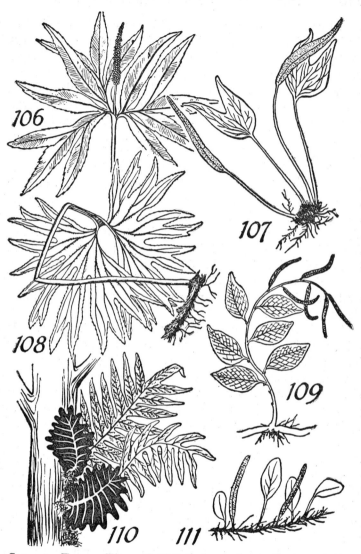

STRANGE FERNS: Fig. 106. *Helminthostachys zeylanica*; Fig. 107. *Cheiropleuria bicuspis*; Fig. 108. *Dipteris conjugata*; Fig. 109. *Photinopteris speciosa*; Fig. 110. *Drynaria quercifolia*; Fig. 111. *Drymoglossum carnosum*.

literally means "to watch the sky," because of its terminal erect fruiting spikes. This name is applied to various other plants in Malaysia and in the Philippines, including one tree (*Polyscias nodosa*) characterized by its great terminal inflorescences extending above its large decompound leaves, and to that species of banana (*Musa uranoscopos*) characterized by its erect rather than pendulous bunches of fruit. In this case, in fact, the specific name of Greek origin was a straight translation of the Malay one meaning "watcher of the heavens." At higher altitudes more and more different species will appear, including the stiff scrambling *Gleichenia* which often forms dense masses or thickets, and on ridges in the constantly wet mossy forests the stiff erect scaly stems of *Oleandra*, with narrow entire leaves. Here also the very striking *Dipteris conjugata*, fig. 108, will be observed in open places, with its large broad two-lobed fronds, the lobes again with narrow lobules. While the number of different fern species at low altitudes may be rather limited, they become overwhelming in the wet mossy forests above an altitude of about 3,000 feet.

Additional strange forms may be observed, some at low, others at higher altitudes; some are terrestrial, others epiphytic. Thus at low altitudes occur the strange *Polypodium sinuosum*, fig. 94, mentioned in the discussion in Chapter 7 under symbiosis, the equally strange *Antrophyum plantagineum*, fig. 103, growing on boulders and trees, and two other epiphytes with entire leaves—the very abundant *Cyclophorus lanceolatus*, fig. 104, and the smaller but somewhat similar *Drymoglossum carnosum*, fig. 111. At higher altitudes and always in the primary forests is the widely distributed *Cheiropleuria bicuspis*, fig. 107, with strikingly dimorphic fronds, as well as *Photinopteris speciosa*, fig. 109, and again the utterly strange *Lecanopteris carnosa*, fig. 95, its basal parts very greatly enlarged, irregularly tunneled, and the abode of ants. Often along small streams in forested regions one may literally have to clear a path through quantities of the gigantic *Angiopteris evecta* and allied species; the great fronds grow up to twelve to fifteen feet in length, with stalks as thick or thicker than one's wrist.

ORCHIDS

To the layman, the word "orchid" connotes plants with strange-shaped, often brilliantly colored, frequently large attractive flowers, or very bizarre ones, grown in conservatories at home by orchid fanciers, or commercially raised and sold at high prices by florists. The orchid cult is an old one, and literally the forests of all tropical countries have been searched by generation after generation of orchid hunters intent on discovering new or strange forms. While many of the commercial orchids originated in tropical America, others came from India, Burma, the Malaysian region, and other parts of the world. The total number of different orchid species in the vast insular region between Asia and Australia is very great, for this archipelago supports one of the very richest orchid floras of the world. A considerable number of our most attractive cultivated species came from this region, particularly representatives of such genera as *Dendrobium, Phalænopsis, Vanda, Cymbidium, Eria, Renanthera, Cypripedium,* and others.

However, the casual visitor may be disappointed in the paucity of showy orchids that he may encounter, for generally speaking, at least at low altitudes, the plants occur as widely scattered individuals, not in masses, and not infrequently they are perched on the branches of some giant tree where they may escape detection. Again it should be remembered that all easily accessible regions have been searched over and over again by professional orchid hunters for the last century. The net result is that species which might previously have been fairly abundant are now relatively scarce. Thus only once in the many years spent in Philippine exploration, did I ever see a tree bedecked with numerous specimens of the very attractive *dápo maripósa* (*Phalænopsis amabilis*) in full flower, and this happened to be on a remote small island that apparently had never been visited by a professional orchid collector. Single shipments of 10,000 or more mature *Phalænopsis* plants from the Philippines are matters of record, and one shipment said to contain 40,000 was reported in the early decades of the present century, which throws some light on the extent of the orchid-collecting industry.

The total number of known species of orchids for the region approximates 5,000, which is in very striking contrast to their paucity in such a large area as North America north of Mexico, where the entire orchid flora is limited to about 150 different kinds. All of our more northern native species are terrestrial, that is, growing on the ground, but some epiphytic forms occur in the southeastern United States. In the tropics, while terrestrial species do occur in considerable numbers, where they grow on the forest floor and a few even in the open grass-lands, the great majority of the different forms, and most of those with large showy flowers, occur only as epiphytes. There are no parasites among the orchids.

In the tropics, as in temperate regions, there are a certain number of orchid species that are saprophytes. Such plants lack chlorophyll, being white, yellowish, or brown, and for their existence are entirely dependent on decaying vegetable matter present in the soil or on the forest floor. Most of these saprophytic plants are small, and they are all leafless, but among the orchids the coarse clambering *Galeola* may have stems up to an inch in diameter and ten to twelve feet long. Representatives of this genus occur only in damp forests, where they thrive on the decayed remains of large trees. They are not common.

In contrast to the orchids with large showy flowers is a very much larger number of species that, as distinguished from commercial types, are often spoken of as "botanical orchids." Here fortunately many orchid fanciers in the past introduced hundreds of species into greenhouse cultivation, because of their interest in the family as a whole. Many individuals have made it their life work to name and describe these species, so that our knowledge of these strange plants has been enriched by their having had access to living plants. Much additional information has also been derived from an intensive study of dried specimens. Very much still remains to be accomplished, for we are still far from having a thorough knowledge of the rich orchid flora of the Old World tropics.

The range in size of orchid plants is enormous. Many of the

species are small and inconspicuous, others are actually minute, and still others relatively gigantic in size. Thus *Grammatophyllum speciosum* forms great clumps, sometimes even growing on the large branches of mangrove swamp trees, with great stems up to ten or more feet in length and two inches in diameter. Contrasted to this is the small and insignificant *Tæniophyllum*, which may easily be overlooked because it has no stems or leaves and is reduced to its green creeping roots and very short slender pedicels and insignificant flowers. The slender flattened green roots radiate from a definite center and are closely attached to the bark of the trunks and branches of the trees. In the center a few short very slender stalks bear the small and inconspicuous flowers and fruits. The roots function as leaves because of the presence of abundant chlorophyll in them, and also hold the plant firmly in place while at the same time they absorb the necessary moisture and nutrients essential to the growth of the plant. There are all possible gradations in size as between the giant *Grammatophyllum* and the tiny *Tæniophyllum* in Malaysia.

Orchids are widely distributed in nature in the tropics and sometimes appear to thrive under the most adverse conditions, for a few occur immediately back of the seashore, others on the apparently dry branches and trunks of trees in exposed situations, others among the coarse grasses on dry slopes and even in swamps, and still others at high altitudes in the mountains; but the total number of species that will normally be observed at low altitudes is nowhere very great. Much depends on the seasonal distribution of the rainfall. In many mangrove forests, especially those located in regions where long dry seasons prevail, orchids will be practically non-existent, and such species as may occur for the most part will be small and insignificant. Mangrove forests in wetter regions may have a considerable orchid flora; a number of different species may be present, as well as numerous individuals of each species, especially where the mangrove occurs along river-banks. In general, few orchids will be found in second-growth forests, and just as few, except for occasional terrestrial ones, in the primary forests at low altitudes. Most epiphytic orchids that do grow in the primary forests are in the treetops, for they need relatively strong light, and in the interior of the forest the light is always dim.

As one attains some altitude in the primary forests, the orchid situation changes very markedly. At altitudes approximating 3,000 feet one encounters the beginning of the mossy forest; and the higher one ascends, the more pronounced this situation becomes. These are the regions of continuous humidity, for in addition to rain, the forests are constantly drenched for days at a time or night after night by the moisture from low clouds. The moss blanket never dries out, and everywhere the ground, ledges, and trunks and branches of trees are covered with thick layers of dripping wet moss. This is the optimum habitat for a great many different kinds of orchids, and this habitat extends upward to the very limits of vegetation. Here the tree trunks and branches bear great masses of orchids, intermingled with epiphytic ferns and other plants adapted to this particular habitat—in very truth the orchid-hunter's paradise. The number of different species is very great, and the number of individual plants even greater. These orchids of the mossy forest are, for the most part, small-flowered species, but are nevertheless frequently most attractive from the abundance of their flowers. In some species the flowers may be large and showy.

Orchid hunting is a fascinating game, whether one be interested in those species that have large and attractive flowers or those that have small and often really insignificant ones. Should interest lie only in the former, the best results will probably be obtained by enlisting the services of natives who are acquainted with the forest and with what it contains, for they will know where to search for the desired plants. If one's interest extends to the small-flowered species, many living plants can be assembled with little trouble, for fortunately these epiphytic orchids can be gathered, transported with little or no care, and re-established with little trouble near one's base, where they may be observed at leisure as they approach their flowering periods. Even orchids brought to low altitudes from the mossy forest will usually survive for some time if placed in a sheltered spot and provided with sufficient moisture, and such transported plants will usually produce flowers. An orchid plant removed from its host tree is usually very easily established in a new host, as new holding roots are rather quickly developed.

It is rather curious to note that most orchid species other than a few very showy ones and some of economic importance, for one reason or another, do not have fixed local names. Thus in the Philippines everywhere orchids as a group are indicated by the word *dápo,* but this same word is also applied to epiphytic plants representing other families. The word means "to perch," and is thus the equivalent of the technical term epiphyte.

MISCELLANEOUS BOTANICAL PROBLEMS

In the tropics everywhere, and particularly in the inadequately known parts of the great insular region between tropical Asia and Australia, unsolved botanical problems are infinite. At first the observer is overwhelmed by the tremendous development in both genera and species, and the relatively simple matter of identifications and relationships impresses him as an endless task. His special interest may be in natural groups, such as the ferns and fern allies, the grasses, orchids, or even the mosses, lichens, fungi, or algæ, or he may be intrigued with the trees, or with the ornamental or economic plants, or even with the diseases of plants. In each case the problem of identification comes first, for without it further investigations are greatly limited. The best advice that can be given to any individual, no matter what his interest in plants may be, is that he prepare and submit specimens to some botanist familiar with the general flora of the region, or to some specialist, if his interest be confined to a single natural group.

Once the relatively simple problem of identification is solved, the field broadens. If the individual be interested in ecology— that is, the conditions that govern the establishment and extent of various types of plant formations—here naturally the first essential is that he know the names of the plants with which he must deal, whether there be few or many species. If he be interested in plant physiology, he must likewise know the names and the relationships of the plant or plants with which he is concerned. If his interest be in the diseases of plants, he must know the name of the host plant and that of the parasite that causes the damage. Added to these fields is the broad one of morphology, dealing with the form and structure of plants and plant parts, where naturally the importance of the name of the plant is paramount. In all cases, without the generally accepted name of the species involved, comparisons with the published work of others is impossible.

Simple problems are very numerous. Take as an illustration the adaptations by which vines and lianas may reach the tops of the highest trees in the great forests—whether this be by twining, by the development of special roots or special organs in the form of hooked spines, hooked inflorescences, sharp stout retrorse teeth, tendrils, and whether the latter be developed as independent organs or as parts of the leaves or of the inflorescences, and whether they operate by twining about the parts of other plants or adhere by simple terminal discs. Another illustration of the infinite number of problems involved is that appertaining to the special adaptations of fruits and of seeds by which the latter are disseminated. Here often the wind plays an important part, and many fruits and seeds are supplied with wings that insure wide distribution. Individuals operating in the primary forest may observe fluttering down through the quiet air the magnificent winged seeds, six or seven inches across, of *Macrozanonia macrocarpa,* the thin white shining wings in outline suggesting those of an airplane. These come from the large fruits of a high-climbing vine belonging in the squash family (Cucurbitaceæ). Much more frequent, however, are those seeds or small fruits that are supplied with terminal tufts of elongated soft white hairs, as in the milkweed family and many of the Compositæ.

Still another category is that where the fruits or seeds are adapted to dissemination by adhering, through the agency of hooked spines or bristles, or by other means, to the fur of animals and to the clothing of man; it is to this category that many of the ultra-successful weeds belong. The fleshy fruits that are eaten by animals and birds insure the wide distribution of those seeds that pass through the intestines unharmed, but here again, occasionally one will note certain very glutinous or sticky fruits that are clearly distributed by their ability to adhere to the feathers of migratory birds.

Along the seashore will be found a considerable number of fruits and seeds in the detritus cast up by the sea. In most cases these will be found to have certain definite adaptations for distribution through the medium of ocean currents, often through the development of corky tissue in either the fruits or the seeds; but some seeds, so distributed, have definite air spaces that assist in making the seeds buoyant. An adaptation here is that seeds that are widely distributed by floating are impervious to sea water and retain their viability over long periods of time while floating; and yet once on land and in contact with fresh water, the seeds will germinate. These adaptations are highly specialized ones, and in general apply only to the fruits or seeds of those plants that grow naturally along the seashores and along the banks of tidal streams.

8
Weeds and Their Significance

Wherever one observes the vegetation in the settled areas of the tropics, he will note a rather characteristic flora typical of those areas where the forests have been destroyed in connection with primitive or advanced agricultural activities, fallow fields, roadside gardens, and waste lands in and near settlements. This is a mixed flora made up very largely of weeds and weedy plants, including various grasses and sedges. In the vast stretches of the Pacific region most of the species are not natives of the regions they now occupy but have been introduced, for the most part accidentally by man and largely within the historic period; this statement applies to both Polynesia and Malaysia. Wherever man travels and establishes permanent settlements, these weeds follow him.

If the observer be a botanist reasonably familiar with any tropical flora, he will at once recognize very many of the individual species, for generally speaking the weeds of one tropical region are the same as those in other tropical countries, no matter where located. A knowledge of the weeds of temperate regions will be of little assistance when it comes to considering their tropical prototypes, for the tropical weeds are in large part different from those of temperate zones. Few of these weeds, no matter how dominant or how common they may be, have even reasonably fixed native names—evidence, if further evidence be needed, that most of them are introduced rather than native species.

MAN-MADE HABITATS FOR WEEDS

In the humid tropics it is only in those areas where the original vegetation has been radically altered by the activities of man—and this usually takes the form of destruction of the original forest vegetation to clear the land for agricultural purposes—that any considerable number of weeds will be noted. These weeds originally developed as component parts of the floras of naturally open lands, for they are heliophytes, or "sun plants," as contrasted to the characteristic species that one notes in forested regions, which are naturally adapted to growth in partial or dense shade. Without strong direct sunlight weeds cannot thrive, and hence the introduced ones are rarely found within the forests, except along trails and in open places. Wherever natural reforestation develops on abandoned agricultural lands, the weed flora disappears.

WEEDS MOSTLY INTRODUCED PLANTS

In any given area, weeds once introduced and established tend to become dominant and to persist as long as open rather than forested conditions prevail. Some are definitely ruderal, in that they thrive best in places where the soil is more or less regularly disturbed, as in the preparation of areas for planting crops; but others will persist in competition with native elements as long as the areas remain even in part deforested. Thus, in final analysis, the presence and the persistence of characteristic weeds in tropical regions that were originally forested is dependent on man and his activities; for the very presence of these weed species in regions geographically very remote from their original homes is actually due to man as a most efficient agent in the dissemination of plants. Man long has been, and still is, the most disturbing agent in relation to the alterations of original floras, wherever he has become a dominant factor.

The places where this characteristic weed flora occurs are largely lands devoted to agriculture of one type or another, open places where the soil is disturbed from time to time, fallow fields, coconut plantations where the palms are not too closely planted, and along trails and roads, waste places in and near settlements, old garden sites, and paddy lands.

Throughout the vast undisturbed forest areas of the tropics, weeds are absent. But once the forest cover is removed by any agency and for any purpose, the land normally becomes quickly occupied by other plants. This incoming vegetation may consist of very coarse native grasses, dense thickets of native bamboos, or seedlings of quick-growing shrubs and trees; and these types of vegetation are distinctly characteristic of regions wherein weeds have not been introduced. But wherever the most primitive agriculture is practiced it will be noted that, once aggressive weeds have been introduced, these usually annual or perennial herbs take the lead in occupying the land and become a factor with which to be reckoned. These in turn will be wholly or partly eliminated wherever the characteristic second-growth forests, bamboo thickets, or coarse native grasses become established. Where did they come from, when and how were they introduced, and why are they so aggressive?

TEMPERATE-ZONE AND TROPICAL WEEDS

In the temperate regions of North America most of the several hundred species of weeds are natives of Europe and of Asia, with occasionally a very aggressive one that originated in warmer regions but has been able to adapt itself to the more severe climatic conditions in the north temperate zone. The great centers of origin of tropical weeds are definitely Mexico and parts of South America, especially eastern Brazil, in the New World, and parts of India and Africa in the Old World; that is, those areas within the tropics with a limited rainfall, where an open type of vegetation has developed rather than a dense tropical forest. Naturally areas in which the practice of agriculture is an ancient art form a part of the picture.

WEEDS ESSENTIALLY MAN-DISTRIBUTED

A century or so ago it was generally assumed that many weeds owed their wide distribution and almost universal occurrence in the open lands of the tropics to the ability of these plants to spread over vast stretches of land and water unaided, but the evidence now available is opposed to this idea. It has become increasingly clear that man is, and long has been, the agency by which weeds, native of one country or one continent, have been introduced into virgin territory. Being aggressive by nature, having very viable seeds, producing their seeds in great abundance, often adapted to growth under rather adverse conditions, and usually having special adaptations for the dissemination of their seeds by the wind, by man, and by animals, they quickly spread and rapidly become dominant once they are introduced into a new and open region.

Eighty years ago Seemann, in discussing the weeds of Fiji, suggested that the Polynesian islands acted as a bridge by means of which the weeds of the New World reached the Old World, and vice versa. In 1865 he considered sixty-four species that were then troublesome weeds in Fiji; most of them were very widely diffused, but forty-eight were common to America and only sixteen were strictly confined to the Old World. This bridge assumption is definitely not correct unless one includes the unconscious intervention of man, for it was unquestionably the early European explorers, landing here and there, and later European agriculturists settling in various places, who were the prime factors in this episode of plant introduction. Nearly 250 years passed after the European explorers reached the Pacific basin before botanical collections were made in the region, the first botanical field work being accomplished in and following 1768; and all of the early and fairly extensive botanical collections from all parts of Polynesia made between 1768 and 1780 yielded only thirty-seven species that could by any possibility be classed as weeds.

We may safely assume that some of the relatively few weeds of Old World origin were of early introduction here, having reached Polynesia via Malaysia through the migrating Polynesian peoples themselves, for most of their few cultivated food plants came from the Malayan region. The weeds now known to be of American origin must have reached Polynesia through the various early European exploring expeditions. Here, as elsewhere in the world, once a new weed is introduced and naturalized, it usually spread with remarkable speed within the limits of any single island or land mass; its extension of range to neighboring or remote islands depended on contact through man in connection with his voyages.

PAUCITY OF DEFINITE NATIVE NAMES FOR WEEDS

It is a noteworthy fact that, generally speaking, weeds in the Old World tropics, whether in Malaysia or in Polynesia, have few fixed native names. This is all the more remarkable because these plants are not infrequently the most abundant and the most widely distributed species in the settled areas. In contrast to the introduced weeds, most of the strictly native species, even those of the forest, have reasonably fixed native names. This simple observation provides supporting evidence to the idea that most of the weeds in all of Malaysia and Polynesia are of relatively recent introduction; that is, within the past few hundred years or since the European peoples commenced to explore and colonize the regions under consideration.

HISTORICAL DATA

The activities of the Spaniards in colonizing the Marianas, Caroline, and Philippine Islands beginning about the middle of the sixteenth century resulted in the direct introduction of a great many weeds from Mexico into these islands through the intermediary of the Acapulco-Manila galleons, for this old trade route was active from the early days of Spanish occupation of the Philippines until Mexican independence, covering a period of roughly 250 years. In earlier days there were, of course, the extensive contacts between the more advanced peoples of India, who colonized and controlled much of the Malay Archipelago, and who, in connection with their introduction of the agricultural and ornamental plants characteristic of India, also inadvertently introduced many Asiatic weeds.

RAPIDITY OF SPREAD

Just how rapidly a weed will spread, once introduced, is indicated by a single example. *Borreria lævis,* ubiquitous in tropical

WEEDS AND THEIR SIGNIFICANCE 123

America, apparently did not reach the Old World until after the beginning of the present century. Its first published record was from Java, where it was said to be common in 1924; and yet between the years 1925 and 1929 specimens were actually collected from the very widely separated Sumatra, Singapore, the Philippines, New Britain, New Guinea, and Samoa—new records for all these areas—and it has more recently turned up in Fiji and on Fanning Island. Representatives of such American genera as *Hyptis, Elephantopus, Eichhornia, Waltheria, Mikania, Erechtites, Spilanthes, Clitorea, Lochnera, Synedrella, Cassia, Mimosa, Ageratum, Malvastrum, Sida, Stachytarpheta, Cosmos,* and other characteristic weeds, once introduced into a new region, probably extended their ranges about as fast as the *Borreria* above mentioned.

NAPOLEONIC AMBITIONS OF CERTAIN INTRODUCED SPECIES

One Netherlands Indies botanist mentioned the "Napoleonic ambitions" of certain American weeds of Brazilian origin, once introduced into Java; but certain Old World weeds, once introduced into tropical America, have shown similar ambitions, such as Bermuda-grass (*Cynodon dactylon*), *Eleusine indica, Eclipta alba,* and many others. After all, the matter of "Napoleonic ambitions" of any weed, once introduced into a new country, depends on whether or not the conditions in its new home are favorable to its perpetuation, and on the special adaptations for the dissemination of its seeds. We have the same situation in temperate regions, for in many countries introduced weeds are dominant, such as the very numerous European ones in North America, to mention only such aggressive examples as the dandelion (*Taraxacum officinale*), white-weed or "daisy" (*Chrysanthemum leucanthemum*), and devil's-paintbrush (*Hieracium aurantiacum*), or the Japanese honeysuckle (*Lonicera japonica*); and in other countries the American *Erigeron canadense* in Europe and prickly-pear (*Opuntia*) in Australia. Wherever aggressive aliens become established, they will persist as long as favorable growth conditions prevail. Many of them are more vigorous in their new homes than in the lands wherein they originated, and much more vigorous than the native species with which they compete for space. Perhaps this has been due, at least in some cases, to the fact that in their introduction they were unaccompanied by specific plant diseases and insect pests that tended to keep them within bounds in their original homelands.

This same principle applies to various introduced and naturalized insects, birds, and animals, as well as plants, as witnessed by the success of the Japanese beetle, English sparrow, and European starling in eastern North America, the European rabbit in Australia, and the American muskrat in Europe.

METHODS OF INTRODUCTION

Weed seeds may originally have been transported from one country or one continent to another in packing material, in soil in which living plants were being shipped, by being inadvertently mixed with seeds of economic plants, or merely by adhering to the clothing of man or to the hairs of domestic animals. The wind plays its part, for many seeds have tufts of hairs or wings by which they are wind-distributed; but this wind factor is effective only over relatively limited distances, for wind-blown seeds in general cannot traverse wide expanses of the sea such as the Atlantic, the Pacific, and the Indian oceans. Wind distribution is, however, very effective once a plant is introduced into a new region. Birds and animals play their part, and occasionally fresh water is involved, as with the water-hyacinth; and yet in the case of the latter plant man was the prime factor, for he originally distributed this South American species as an ornamental plant.

RAPIDITY OF WEED FLORA DEVELOPMENT

How rapidly weeds may appear in a virgin area is well illustrated by the case of Christmas Island, in the Indian Ocean, south of western Java. Up to 1888, before any settlers established themselves there, no weeds were present. Two years after the first settler arrived, four weeds were recorded; in 1897, twelve, and in 1904, thirty—a record of only eighteen years following the establishment of the first plantation on the island. If weed seeds had accidentally reached this island before 1888, the plants could not have established themselves for the simple reason that the entire land surface was covered with a dense forest growth.

UNIVERSAL OCCURRENCE OF PANTROPIC WEEDS

For all practical purposes, a weed flora of any long-settled area in the islands of the Pacific or in any part of the Malay Archipelago would also be a weed flora of Porto Rico, Cuba, Panama, Siam, Burma, Ceylon, India, most parts of tropical Africa, or even such a city as Cairo, Egypt. One cannot very well encompass a weed flora in such a work as this; about all that can legitimately be done is briefly to indicate a few of the more common and widely distributed types.

SOME COMMON AND CHARACTERISTIC WEEDS

The number of natural families of plants represented in this flora is not great, but they include, other than the grasses and the sedges, such monocotyledonous groups as the Cannaceæ (*Canna indica*) of American origin, the so-called Indian shot or *tikas-tikas*, with its red flowers, some representatives of the ginger family, and a fairly large number of genera and species belonging to the Commelinaceæ, which are mostly more or less succulent herbs, usually with blue flowers. *Commelina benghalensis*, fig. 114, *C. diffusa*, fig. 117, and *Cyanotis cristata*, fig. 116, are often abundant; incidentally, most or all representatives of this particular family may safely be eaten as substitutes for spinach.

In the dicotyledonous families, the number of weeds is very large, for several hundred different species are involved, and nearly all of these are of more or less universal distribution in all tropical countries. In the nettle family one notes in abundance in the settled areas, usually in damp places and on old walls, the small, somewhat fleshy, so-called gunpowder plant, *Pilea microphylla,* and in open places the common *Pouzolzia zeylanica.* The Chenopodiaceæ is poorly represented in the tropics, but one ubiquitous species is the aromatic *Chenopodium ambrosioides,* fig. 124. The closely allied Amaranthaceæ is rich in weeds, including representatives of such genera as *Achyranthes, Ærua, Alternanthera, Amaranthus, Celosia, Cyathula, Pupalia,* and *Gomphrena.* Of these, the genus *Amaranthus* is best represented, two of the very commonest species being *A. viridis,* fig. 118, and the much coarser, somewhat spiny *A. spinosus,* fig. 120. *Celosia argentea,* fig. 113, may be abundant in fallow fields and waste places; its shining, usually pale pink inflorescences are characteristic. *Alternanthera sessilis,* fig. 112, a prostrate plant with small white axillary heads, and the much coarser, sprawling or erect *Achyranthes aspera,* fig. 115, dominate certain habitats; both are usually abundant. It may be worthy of note that most of the representatives of this family may be utilized as food by cooking the tender parts of the stems and the young leaves. The common purslane, *Portulaca oleracea,* fig. 122, with its fleshy stems and leaves and small yellow flowers, is abundant in and near settlements, and is just as much of a weed in the tropics as it is in the temperate zone; it too may be eaten either raw or cooked. One very common *Oxalis* with yellow flowers and acid leaves is *O. repens,* fig. 140, the tropical representative of the group to which our northern *O. corniculata* belongs. Two ubiquitous weeds in the caper family (Capparidaceæ) are *Polanisia icosandra,* fig. 121, with yellow flowers, and *Gynandropsis gynandra,* fig. 119, with pink flowers. The poppy family is represented by the Mexican poppy, *Argemone mexicana,* fig. 123, with its rather large yellow flowers, yellow sap, and thistle-like, usually mottled leaves.

The bean family or Leguminosæ is remarkably rich in weeds, with various species in such genera as *Æschynomene, Alysicarpus, Cassia, Crotalaria, Desmodium, Flemingia, Indigofera, Sesbania, Zornia,* and others. *Cassia alata,* fig. 127, is a coarse, erect, somewhat shrubby plant with large yellow flowers and four-winged pods, a native of Mexico; *C. tora,* fig. 134, is another species of American origin, as is *C. occidentalis,* fig. 125.

WEEDS AND THEIR SIGNIFICANCE

These aliens in the Old World, particularly *Cassia tora,* fig. 134, may be so abundant as to occupy vacant lots and waste places in and about towns to the almost entire exclusion of other kinds of plants. Another genus rich in weeds is *Crotalaria;* some of the species are of American, others of Old World origin. Thus *Crotalaria mucronata,* fig. 135, *C. retusa,* fig. 136, *C. incana,* fig. 137, and *C. quinquefolia,* fig. 139, all with yellow flowers and all with inflated pods, are almost certain to be noted, as well as the blue-flowered *Crotalaria verrucosa.* The sensitive-plant (*Mimosa pudica,* fig. 138) is of American origin but is now thoroughly at home throughout the settled parts of the Old World tropics. It is conspicuous because of its small heads of pink flowers and further because of the fact that when even lightly touched the leaves respond by collapsing immediately, and after a short rest gradually resume their normal positions. A very common vine in this group is the slender crab's-eye vine, widely known as *sága* (*Abrus precatorius,* fig. 129); its very hard shining seeds, half black and half bright red, are widely used as beads. *Clitoria ternatea,* fig. 141, another vine of American origin but curiously first named from material originating in the Old World, is characterized by its strangely shaped, usually blue, fairly large flowers an inch long or more, and its flat pods.

One species of the soapberry family (Sapindaceæ) is rather striking in spite of its insignificant flowers. This is *Cardiospermum halicacabum,* fig. 147, a slender, tendril-bearing vine with compound leaves; its fruits are inflated, obovoid, about an inch long, each of the hard black seeds with a white heart-shaped aril at the base, whence the generic name, which means "heart" and "seed." In the Euphorbiaceæ there are a considerable number of *Euphorbia* species, some of these the very commonest of weeds; all of them are characterized by milky sap. The most common and most ubiquitous species is *Euphorbia hirta,* fig. 143, with leaves usually somewhat purplish-blotched in the middle; species with much smaller leaves are *E. thymifolia* and *E. prostrata. Euphorbia heterophylla,* fig. 142, was originally disseminated as an ornamental species, being of American origin, but is now a common weed. It is an erect plant, with basal parts of the leaves below the flowers bright red. In the same family may be mentioned *Acalypha indica*, fig. 146, various herbaceous species of *Phyllanthus*, such as *P. niruri*, fig. 144, the very similar *P. urinaria*, and the coarse castor-oil-plant, *Ricinus communis*, fig. 145.

Three closely allied families, the Malvaceæ, Tiliaceæ, and Sterculiaceæ, like the Leguminosæ, are also rich in weeds. Several representatives of the genus *Triumfetta* are frequently very abundant, particularly *T. bartramia*, fig. 149, and the very similar *T. semitriloba*; they are successful immigrants because of the ease by which their seeds are distributed. The small fruits are supplied with small hooked spines by which they adhere to one's clothing and to the fur of animals. Three species of *Corchorus*, a genus in the same family as *Triumfetta*, the Tiliaceæ, are ubiquitous, all characterized by peculiar descending tail-like appendages near the leaf bases. They are *Corchorus olitorius*, fig. 151, with elongated fruits, its tender parts utilized for food, and the very similar *C. capsularis*, which, however, has globose fruits—both strictly erect herbs—and *C. acutangulus*, a more or less prostrate herb with elongated pods. Of these, *C. capsularis* is the source of the commercial fiber jute, which is prepared from the bast of that species. Outstanding in the mallow family (Malvaceæ) as tropical weeds are various representatives of the genus *Sida*, all plants with yellow flowers, such as *S. rhombifolia*, fig. 157, *S. acuta*, fig. 158, and *S. cordifolia*, fig. 154. Another common yellow-flowered weed, with flowers larger than in *Sida*, is *Abutilon indicum*, fig. 148, while *Malvastrum coromandelinum*, fig. 150, is another weed of American origin first named from Old World material. *Abelmoschus moschatus*, fig. 155, is a coarser plant with much larger yellow flowers than the species above mentioned; its seeds, when rubbed or crushed, have a striking musk-like odor, whence its specific name. Another very common and variable species is *Urena lobata*, fig. 153, with pink flowers. *Hibiscus surattensis*, fig. 159, is a sprawling plant, its stems with short spines, the flowers yellow with a purple center. In the third family in this group, the Sterculiaceæ, two typical weeds are *Waltheria americana*, fig. 152, and *Melochia corchorifolia*, fig. 156, the former with small yellow flowers and the latter with somewhat larger pink or whitish ones.

In the milkweed family (Asclepiadaceæ), the most common weed is *Asclepias curassavica,* fig. 162, of American origin, and conspicuous because of its red and yellow flowers. In the allied family, the Apocynaceæ, which likewise is characterized by the presence of milky sap, the attractive *Catharanthus roseus,* fig. 163, is often abundant; this, like the preceding species, is also of American origin, probably first distributed as an ornamental plant. Its flowers vary from white to pink or red, or white with a pink or red center, and in size from one and one-half to two inches in diameter. Where it has been introduced, it finds the sandy seashores a favorable habitat. A frequently common vine of the passion-flower family is *Passiflora fœtida,* fig. 160, its flowers and fruits surrounded by very finely divided bracts; this is of American origin.

The morning-glory family (Convolvulaceæ) is particularly well represented by various species of *Ipomœa, Merremia, Hewettia, Quamoclit, Operculina,* and *Calonyction.* They are all vines, although some of the smaller ones may be prostrate or creeping; they are more characteristic of fence rows and thickets than of the open country, although a considerable number occur in grass-lands. *Ipomœa pes-tigridis* has hairy, five- to nine-lobed leaves and white flowers, *I. cairica* has smooth five- or seven-parted leaves and pale lavender flowers, while *I. digitata* has large lobed smooth leaves and large pink flowers, and *I. hederacea* has hairy three-lobed leaves and very attractive pale blue flowers. In the closely allied genus *Merremia,* the flowers are mostly yellow, as in *M. hirta, M. gemella,* and *M. emarginata,* but in the very narrow-leaved *M. hastata* they are pale yellowish or nearly white with a purple center, and in *M. umbellata* they are larger than in the other species, and white. *Merremia vitifolia,* with normally five-lobed hairy leaves, has large yellow flowers. The *Quamoclit* species are the cypress-vine (*Q. pennata,* fig. 165), with very finely divided leaves, and *Q. coccinea,* fig. 164, with ovate, entire or broadly three- to five-lobed leaves; both are of American origin, and both have deep-red flowers. The *Calonyction* flowers are very large, white, and open only at night, the common representative being the moonflower, *C. aculeatum.*

In the borage family (Boraginaceæ), the most striking of the weed species is the Brazilian *Heliotropium indicum,* fig. 161, and in the closely allied verbena family (Verbenaceæ), *Stachytarpheta jamaicensis,* fig. 168, a coarse erect herb with slender spikes of scattered blue flowers, is usually common. Even more abundant in open places is the prostrate *Lippia nodiflora,* fig. 166, with small heads of pink flowers.

Another family in this alliance, also noted for the large number of weeds, is the mint family (Labiatæ). In this group the stems are usually square and the plants are often distinctly aromatic when crushed. Characteristic weed genera are *Moschosma, Ocimum, Hyptis, Anisomeles, Leucas,* and *Leonurus.* Probably the most common weed in the settled areas of the Orient is *Hyptis suaveolens,* fig. 171, but *H. capitata, H. brevipes,* and *H. spicigera* are often abundant; these are all plants of American origin. If one is interested in observing adaptations by which seeds are disseminated, he should walk through a growth of *Hyptis suaveolens* when the plants are wet, and note how the small fruits adhere to his clothing by a white sticky mucous-like substance that covers the outside of small seeds; yet when the seeds are dry, there is no indication whatever of this character. *Anisomeles indica,* fig. 167, a coarse, rather softly pubescent plant, is common, while in the genus *Leucas* a number of weeds occur, such as *L. aspera,* fig. 181, *L. lavandulifolia,* fig. 178, and *L. zeylanica,* fig. 179; all of these have white flowers. *Leonurus sibiricus,* fig. 170, is a rather tall coarse plant with finely divided leaves and reddish flowers.

Closely allied to the mint family is the Scrophulariaceæ. The various species, like the mints, have usually very irregular flowers, but the fruits are many-seeded capsules, while the stems are usually round rather than square; few of the species are aromatic. Genera that include weeds are *Bacopa, Lindernia, Limnophila, Torenia, Centranthera, Ilysanthes, Dopatrium,* and above all *Scoparia.* However, in this group the plants are mostly small and do not tend to become dominant as do various representatives of the mint family. In the potato family one of the very common weeds is *Solanum nigrum,* fig. 169, its vegetative parts widely used as a substitute for spinach in all tropical countries. Also common are various coarser representatives of the same genus, several species of *Physalis,* such as *P. lanceifolia,* fig. 172, *P. minima,* fig. 173, and *P. peruviana,* fig. 174, with small tomato-like fruits surrounded by a loose husk. Often one will note in abundance the wild form of the common tomato, *Lycopersicum esculentum,* its red fruits less than one-half inch in diameter. Locally a jimson-weed will be conspicuous, *Datura metel,* fig. 177, a coarse plant with large leaves and very large white or sometimes purplish, fragrant flowers. In this same family is the Chile pepper (*Capsicum frutescens*), which is frequently spontaneous. It is characterized by its small red fruits which are exceedingly peppery when tasted. In the limited group of families characterized by markedly irregular flowers is the Acanthaceæ, but here more of the species are cultivated for ornamental purposes. However, some of the introduced species have become thoroughly naturalized, such as the attractive *Thunbergia alata,* fig. 175, native of Madagascar, with its prominently winged leafstalks and rather large yellow flowers with white or dark purple centers. Various representatives of other genera, some apparently native, some introduced from America, rate as weeds in such groups as *Elytraria, Justicia, Hygrophila,* and *Blechum.*

Most of the tropical Rubiaceæ are woody plants, but some herbs occur, and some of these, such as *Dentella repens,* several species of *Hedyotis* (*Oldenlandia*), and *Borreria,* are definitely weeds; some of the species are abundant. They are mostly small plants with strictly opposite leaves and small regular flowers.

Standing at the top of all tropical families of plants that have produced the greatest number of weeds is the sunflower family (Compositæ). The success of its representatives, once introduced into a new area, is apparently due partly to the various special adaptations by which the small fruits are distributed. Many of the Old World weeds in this very strongly characterized family originated in tropical America, and most of them were introduced into the Old World early in the history of exploration. *Vernonia cinerea,* fig. 180, and *V. patula,* both with purple flowers, are common in waste places. The light blue *Ageratum conyzoides,* fig. 176, and the pink *Emilia sonchifolia,* fig. 184, are abundant. White- and purple-flowered forms of *Elephantopus* occur, the most common being *E. scaber,* fig. 186. Everywhere in and near settlements will be noted such species as the yellow-flowered *Synedrella nodiflora,* fig. 183, a coarse species of *Erigeron* (*E. sumatrensis*) with small dirty white flowers, and the white-flowered *Eclipta alba.* The ubiquitous beggar-ticks, *Bidens pilosa,* fig. 187, with white flowers, and the closely allied but much coarser *Cosmos caudatus,* fig. 188, with pink or pale purple flowers, as well as a representative of the American genus *Mikania,* a vine with many small whitish flowers, are usually common. Two species of the American *Erechtites,* both with pink flowers, may be dominant wherever they have been introduced, *E. hieracifolia,* fig. 185A, and *E. valerianæfolia,* fig. 185B. *Adenostemma lavenia,* fig. 182, with white flowers, is more characteristic of rather damp, shaded places. Among common species, also with white flowers, is *Dichrocephala latifolia,* fig. 189. This summary by no means exhausts the list of genera of the Compositæ that have supplied weeds to the Old World tropics; in general their representatives there are accidental introductions from other parts of the world. Other genera involved are *Sphæranthus, Eupatorium, Blumea, Grangea, Artemisia, Centipeda, Sonchus, Crepis, Spilanthes,* and *Tridax.*

Most of the weed species are annual herbs, and most of them have certain adaptations rendering the dissemination of their seeds simple and easy. They are plants that are definitely adapted to growth in the open rather than in shady places. While weeds are plants inimical to agriculture, they are not to be ignored merely because they are customarily found on land that is cultivated or has been in cultivation, or because they are relatively recent arrivals. Everywhere because of their abundance they become conspicuous in their favored habitats.

From a purely botanical standpoint, weeds are perhaps uninteresting. One concerned with locating rare, imperfectly known, or even quite unknown species will ignore them, for, at least if he be a trained botanist, he may realize that most of the species are aliens in the Pacific islands where so many of them now thrive. Yet now being component parts of the various floras, no matter when they were introduced, and no matter how unattractive they may be, they cannot be wholly ignored, for whether natives or aliens, they are successful and will persist as long as conditions are favorable to their growth. After all, much may be learned from the lowly weeds, no matter how much they may be anathematized by the gardener or the agriculturist.

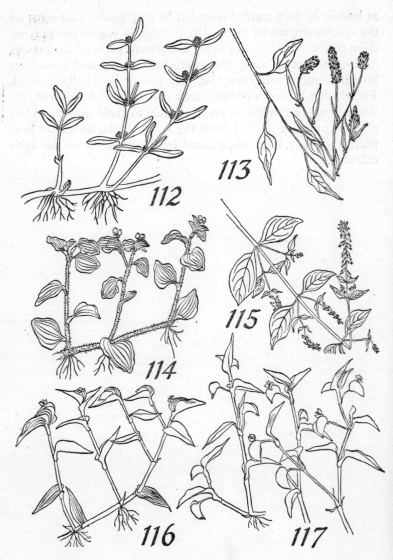

WEEDS: Fig. 112. *Alternanthera sessilis*; Fig. 113. *Celosia argentea*; Fig. 114. *Commelina benghalensis*; Fig. 115. *Achyranthes aspera*; Fig. 116. *Cyanotis cristata*; Fig. 117. *Commelina diffusa*.

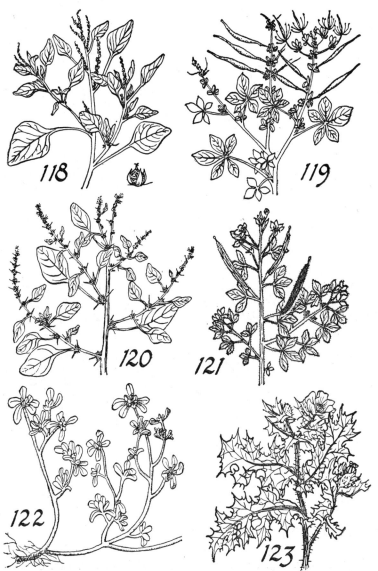

WEEDS: Fig. 118. *Amaranthus viridis*; Fig. 119. *Gynandropsis gynandra*; Fig. 120. *Amaranthus spinosus*; Fig. 121. *Polanisia icosandra*; Fig. 122. *Portulaca oleracea*; Fig. 123. *Argemone mexicana.*

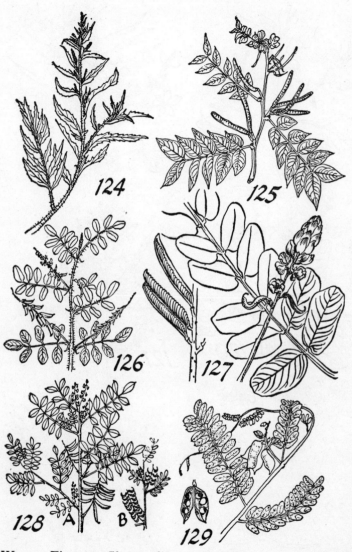

WEEDS: Fig. 124. *Chenopodium ambrosioides*; Fig. 125. *Cassia occidentalis*; Fig. 126. *Indigofera hirsuta*; Fig. 127. *Cassia alata*; Fig. 128. A. *Indigofera tinctoria*, B. *I. suffruticosa*; Fig. 129. *Abrus precatorius*.

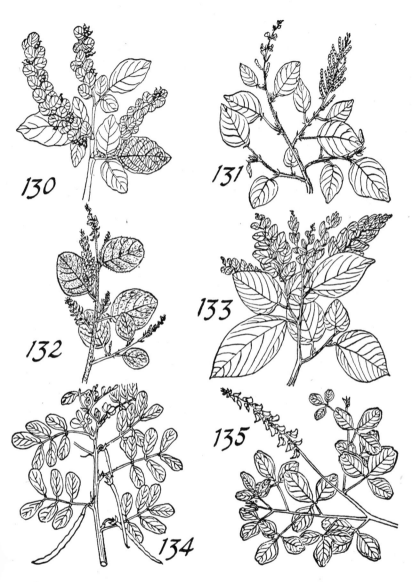

WEEDS: Fig. 130. *Desmodium pulchellum*; Fig. 131. *Desmodium gangeticum*; Fig. 132. *Desmodium velutinum*; Fig. 133. *Flemingia strobilifera*; Fig. 134. *Cassia tora*; Fig. 135. *Crotalaria mucronata.*

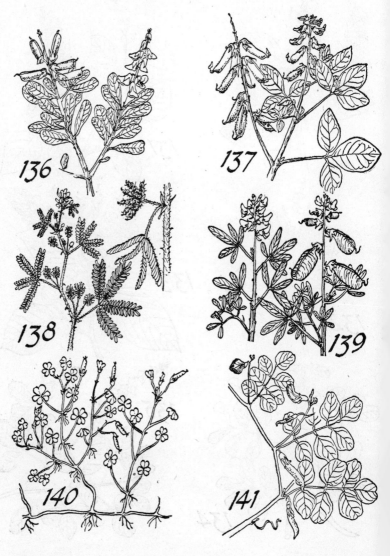

Weeds: Fig. 136. *Crotalaria retusa*; Fig. 137. *Crotalaria incana*; Fig. 138. *Mimosa pudica*; Fig. 139. *Crotalaria quinquefolia*; Fig. 140. *Oxalis repens*; Fig. 141. *Clitorea ternatea*.

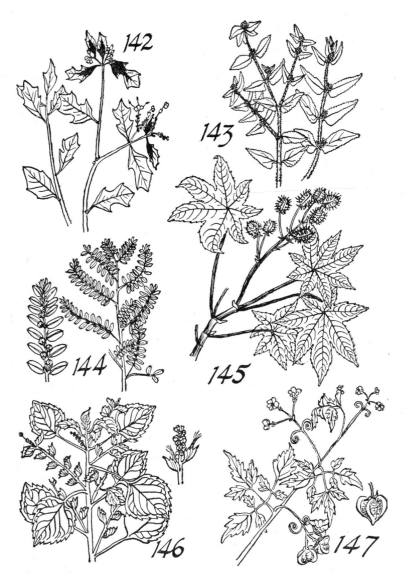

WEEDS: Fig. 142. *Euphorbia heterophylla*; Fig. 143. *Euphorbia hirta*; Fig. 144. *Phyllanthus niruri*; Fig. 145. *Ricinus communis*; Fig. 146. *Acalypha indica*; Fig. 147. *Cardiospermum halicacabum*.

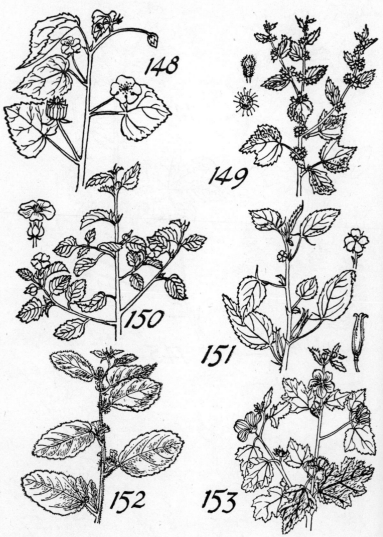

Weeds: Fig. 148. *Abutilon indicum*; Fig. 149. *Triumfetta bartramia*; Fig. 150. *Malvastrum coromandelinum*; Fig. 151. *Corchorus olitorius*; Fig. 152. *Waltheria americana*; Fig. 153. *Urena lobata*.

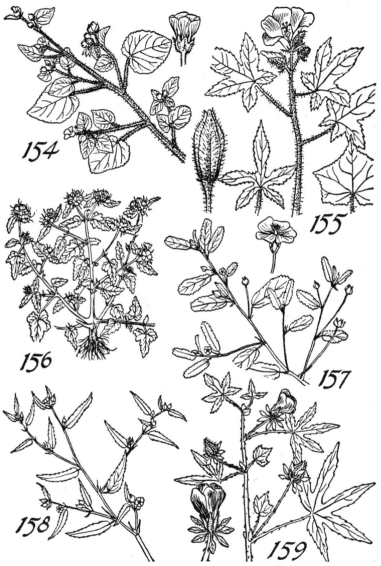

WEEDS: Fig. 154. *Sida cordifolia*; Fig. 155. *Abelmoschus moschatus*; Fig. 156. *Melochia corchorifolia*; Fig. 157. *Sida rhombifolia*; Fig. 158. *Sida acuta*; Fig. 159. *Hibiscus surattensis*.

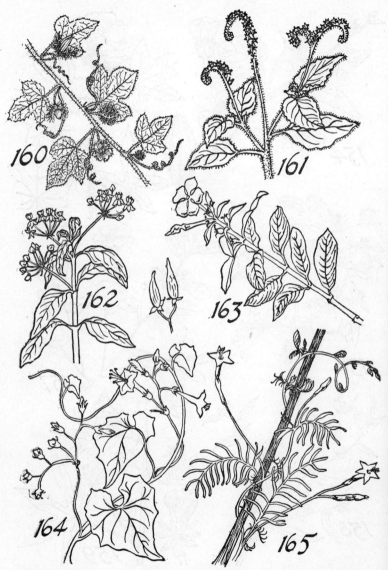

WEEDS: Fig. 160. *Passiflora foetida*; Fig. 161. *Heliotropium indicum*; Fig. 162. *Asclepias curassavica*; Fig. 163. *Catharanthus roseus*; Fig. 164. *Quamoclit coccinea*; Fig. 165. *Quamoclit pennata*.

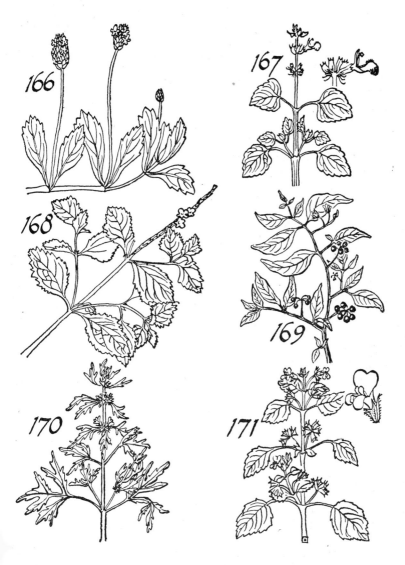

Weeds: Fig. 166. *Lippia nodiflora*; Fig. 167. *Anisomeles indica*; Fig. 168. *Stachytarpheta jamaicensis*; Fig. 169. *Solanum nigrum*; Fig. 170. *Leonurus sibiricus*; Fig. 171. *Hyptis suaveolens*.

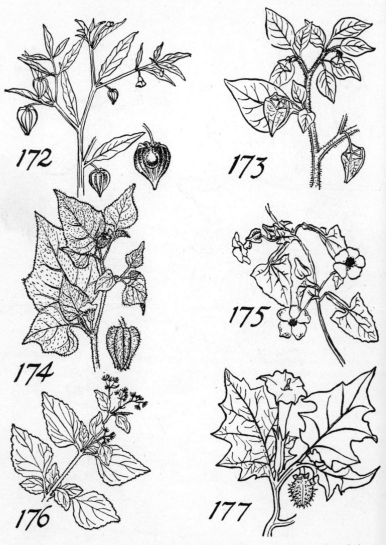

WEEDS: Fig. 172. *Physalis lanceifolia*; Fig. 173. *Physalis minima*
Fig. 174. *Physalis peruviana*; Fig. 175. *Thunbergia alata*; Fig. 176. *Ageratum conyzoides*; Fig. 177. *Datura metel*.

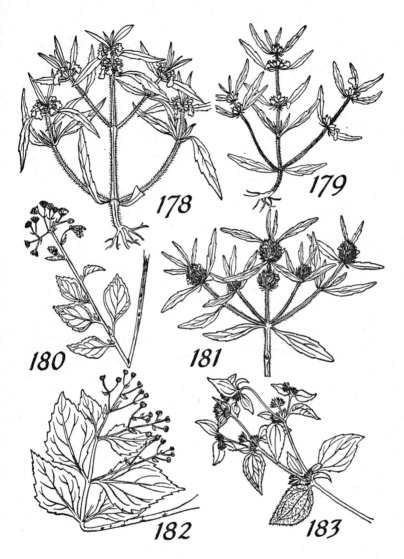

WEEDS: Fig. 178. *Leucas lavandulifolia*; Fig. 179. *Leucas zeylanica*; Fig. 180. *Vernonia cinerea*; Fig. 181. *Leucas aspera*; Fig. 182. *Adenostemma lavenia*; Fig. 183. *Synedrella nodiflora*.

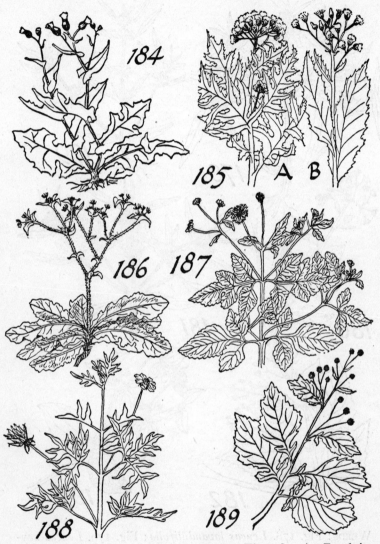

WEEDS: Fig. 184. *Emilia sonchifolia*; Fig. 185. A. *Erechtites hieracifolia*, B. *Erechtitea valerianaefolia*; Fig. 186. *Elephantopus scaber*; Fig. 187. *Bidens pilosa*; Fig. 188. *Cosmos caudatus*; Fig. 189. *Dichrocephala latifolia*.

9

The Cultivated Plants

Excluding many vegetables and a few cereals, the cultivated tropical food plants are, in general, totally different from those with which the resident of a temperate country is familiar. He will recognize such commonly cultivated vegetables as cabbage, Chinese cabbage, radish, garden bean, Lima bean, cowpea, pepper, okra, cucumber, squash, eggplant, carrot, beet, celery, tomato, sweet potato, onion, garlic, peanut, and various others, with varieties of which he is familiar at home. Some standard tropical vegetables and all tropical fruits will be strange to him, except as he knows such imported ones as the banana, pineapple, coconut, and the subtropical citrus fruits, including the orange, lemon, lime, and pomelo.

Merely because one does not know the characteristic tropical food plants is no reason why he should overlook them, for many are excellent and form a welcome addition to the diet. Therefore it is advisable to have an entirely open mind and not to refuse to eat this or that merely because it is unfamiliar and untried.

CEREALS

The only important cereals that are commonly cultivated in the Malaysian region are maize or Indian corn, originally from America, rice (*Oryza sativa*), and in places Italian millet (*Setaria italica*), certain forms of sorghum (*Andropogon sorghum*), and a form of Job's-tears (*Coix lachryma-jobi*) with very thin-walled involucres. Here and there, ragi (*Eleusine corocana*) may be planted. These cereals scarcely need further consideration. To this same natural family that includes all of the cereals, the Gramineæ, also belongs the common sugarcane (*Saccharum officinarum*).

STRICTLY TROPICAL VEGETABLES

There are a number of tropical vegetables that will be strange to residents of temperate lands, some cultivated, some more or less widely naturalized. Among the beans are the hyacinth bean or *batáon* (*Dolichos lablab*, fig. 198), a vine with flat thin pods and white or pink flowers, the immature pods of which may be used quite as one would prepare string beans; this is cultivated, and also occurs wild. The asparagus bean, *kalamísmis* or *ketjeper* (*Psophocarpus tetragonolobus*, fig. 196), also a vine with blue flowers and thick pods, square in cross section and with four thin elongated wings running the entire length of the pod, is not uncommon, and is a very excellent vegetable. The peculiar yam bean, *sínkamas* (*Pachyrrhizus erosus*, fig. 200), is sometimes planted but is fairly widely naturalized. It is a vine with blue flowers, utilized not for its pods or seeds—the latter are actually poisonous—but for its turnip-like root, which is always eaten raw; the white flesh is crisp, juicy, and distinctly refreshing. Here and there will be observed such small-seeded beans as the rice bean (*Phaseolus calcaratus*) and the green gram or *múngos* (*Phaseolus aureus*) with small, almost cylindric pods. The former is a vine with slender stems and yellow flowers, the latter an erect herb usually with dirty yellow or greenish flowers. The chief form of the cowpea, *kátjang* or *sítao* (*Vigna sinensis*), is one with very long, often somewhat flaccid pods up to a foot or more in length. Totally different from these vines and herbs is the pigeon pea or *kádios* (*Cajanus cajan*, fig. 197), which is an erect shrubby plant usually five or six feet high, with somewhat yellowish flowers, narrow leaflets, the pods diagonally depressed between the seeds. The immature beans are eaten like green peas, and the mature beans may also be eaten after thorough cooking. In passing, it is desirable to note that the black seeds of the wild forms of the Lima bean (*Phaseolus lunatus*, fig. 201), known as *patáni*, are very poisonous and should not be eaten until after very long cooking; the poisonous principle is hydrocyanic acid.

In the calla lily family (Araceæ), the one almost universally cultivated food plant is ordinary taro, *tálie* or *gábi* (*Colocasia esculenta*, fig. 190), numerous varieties being recognized; in many respects its much-thickened starchy root takes the place of the potato in the tropics. Even its tender young leaves may be cooked and eaten as greens, but they must be well cooked because of the presence of microscopic, stinging, irritating crystals in their tissues. The allied but larger American yautia (*Xanthosoma violaceum*) has come into cultivation for food in Java and elsewhere in recent years.

In the squash family other than the well-known cucumber and squash, the watermelon, and occasionally forms of the ordinary melon, one may note the wax gourd or *kondól* (*Benincasa hispida*). The bottle gourd or *ópo* (*Lagenaria siceraria*), with white flowers and frequently very large fruits, is not uncommon in cultivation, as well as two species of the so-called dishcloth gourd or loofah, *Luffa acutangula*, fig. 212B, with longitudinally angled fruits, and *L. cylindrica*, fig. 212A, with cylindric fruits, both known as *patóla* and as *tabóbok*. Both have yellow flowers, and their immature fruits are excellent as a cooked vegetable. The bitter cucumber, both cultivated and naturalized, with its strange warty fruits and red flesh, *ampaláya* or *amargóso* (*Momordica charantia*), while actually somewhat bitter, is nevertheless widely used as food.

Several true yams—not to be confused with the so-called yam in the United States, which is a variety of the sweet potato—occur both wild and cultivated. The most important are the goa yam, *túngo*, *túgue*, or *támis* (*Dioscorea esculenta*, fig. 67) and the greater yam or *úbi* (*Dioscorea alata*, fig. 191). The tubers of both are exceedingly variable in shape and size, those of the latter sometimes weighing up to forty pounds. Among tuber-bearing plants, but totally unrelated to the yams, is cassava, tapioca, or *kamóting-káhoy* (*Manihot esculenta*, fig. 205). This is an erect shrub with lobed leaves and large tubers rich in starch. These must be cooked before eating, for when fresh they are distinctly poisonous (especially the bitter cassava) because of the presence of hydrocyanic acid, which is dissipated in cooking. The common sweet potato, known in the Philippines under its Mexican name, *kamóte,* is in general the most widely cultivated tuber plant, and incidentally the tender shoots and young leaves are very widely used as a substitute for spinach.

Even the banana, or better, the form known as plantain, the cooking banana, when boiled, roasted, or fried is widely used as a vegetable, and the top-shaped terminal hanging bud at the end of the flower stalk, widely known as *póso,* is extensively used as a cooked vegetable; some forms are too bitter to be palatable.

Particularly in Java, and wherever the Javanese have settled, tender parts of a great variety of wild plants are used as food, being eaten either raw or cooked; many of these are common weeds. And the sojourner, perhaps short on food, should not forget that the tender buds surrounded by the leafstalks at the tips of the trunks of many palms are not only edible, either cooked or raw, but are an excellent vegetable; see "millionaire's salad," p. 183.

STRICTLY TROPICAL FRUITS

As will be noted from the preceding statement, the majority of the cultivated vegetables in the Old World tropics will be more or less familiar to the visitor from temperate countries. In the case of the tropical fruits, the differences are very great. Apples, pears, peaches, plums, cherries, and most of the small fruits so common in temperate regions are absent, for these species will not grow in the tropics except occasionally here and there at high altitudes where some of them may be planted. Thoroughly familiar because of constant importation into the United States from the tropics will be the banana, but instead of two or three varieties, including the *gros michel* and the *morado* currently imported, literally scores of varieties will be observed. Their fruits vary greatly in size and color, as well as in flavor and other characters, for in Malaysia the banana will, in a way, compare with the apple of temperate regions in the total number of recognizable varieties. The fruits of the wild bananas are usually filled with numerous hard black seeds, the pulp being very scanty. Equally familiar will be some of the sweet and sour oranges, limes, citron, and other citrus fruits, although the horticultural varieties of these will, in general, be different from those with which one may be familiar at home. The most spectacular of these will be the shaddock or pomelo, one of the parent species of our grapefruit with their very large and very thick-skinned fruits. Another familiar fruit will be the pineapple (*Ananas comosus*), also of American origin, but now widely planted elsewhere.

Strictly tropical, and not commonly known in the United States, outside subtropical Florida, is the magnificent mango, in many respects the best of all tropical fruits, provided one encounters the better varieties. In addition to the common mango (*Mangifera indica,* fig. 207), other species of the genus occur as wild plants in the forests, and several other species are cultivated for their edible fruits, including the *binjai* (*M. cæsia*), the *báchang* (*M. fœtida*), the *kwini* (*M. odorata*), and the *lánjut* (*M. lagenifera*). It should, however, be emphasized that, while the fruits of all of these species are edible—although not so palatable as are those of the common mango—the sap and even the vegetative parts of the trees are distinct contact poisons which cause irritating skin eruptions directly comparable to dermatitis caused by contact with our poison ivy or poison oak. It is well to be on guard against any of the mangoes that have a strong turpentine odor. Occasionally individuals allergic to the common mango will develop a characteristic mango rash after eating the fruit, but such cases are not common.

In this same family also belongs the common cashew (*Anacardium occidentale,* fig. 206), with its striking fruit. The seed —the cashew nut of our tables—is borne in the small kidney-shaped terminal part of the fruit, and is eaten only after being boiled or roasted; the sap in the walls surrounding the single seed is distinctly caustic, and contact with it may also cause bad skin eruptions. Also in this family is *libas* (*Spondias pinnata*) and the allied *vi* (*S. cytherea*), trees with pinnate leaves and plum-like fruits, sometimes known as the hog-plum and the vi-apple; these trees are not contact poisons. The former is Malaysian and the latter chiefly Polynesian.

In fairly general cultivation will be found three distinct species of the genus *Annona*—the soursop, *nángka blánda* or *guanábanos* (*A. muricata,* fig. 194), its large fruit covered with scattered, rather soft, short, spine-like projections; the bullocks-heart, *anónas* or *búah nóna* (*A. reticulata,* fig. 195), with somewhat smaller fruits, their surfaces with rather faint five-angled markings or impressions; and the custard-apple, sweetsop, *átis* or *sirikája* (*A. squamosa,* fig. 192), its still smaller fruits with irregular knobby surfaces. All of these are eaten raw, and all contain fairly numerous hard black seeds; they are all natives of tropical America.

Commonly planted and frequently naturalized is also the tropical American *papáya* (*Carica papaya,* fig. 213), its soft trunks usually unbranched, and its melon-like fruits borne on the trunk below the large, prominently lobed leaves. Equally abundant and even more frequently naturalized is the guava (*Psidium guajava,* fig. 216), a small tree with globose to pear-shaped, smooth, medium-sized, very seedy fruits, the flesh often pinkish in color; this is also a native of tropical America. Still another fruit tree of American origin is the sapodilla, *chiko,* or *sáwo manila* (*Achras zapota,* fig. 218), a small or medium-sized tree with milky sap and smooth, brownish, globose or ellipsoid sweet fruits containing several hard shining black seeds.

Occasionally one will find in cultivation two herbaceous vines, both originally from tropical America and both producing edible fruits. The coarse vine with narrowly winged stems, entire leaves, and large yellowish green fruits is the granadilla (*Passiflora quadrangularis*), and the other with smaller globose purple fruits and three-lobed leaves is the passion-fruit (*Passiflora edulis*).

Clearly the majority of the cultivated fruit species of the Malaysian region represent introductions, in either prehistoric or historic times; some of these came originally from tropical Asia, including most of the *Citrus* fruits, and others from tropical America. While it is possible that the common mango represents an introduction from the Asiatic continent, it may have originated in some part of the archipelago, as clearly did the other species of *Mangifera* mentioned above. The jakfruit or *nángka* (*Artocarpus heterophylla*, fig. 85), although now widely distributed in the archipelago and frequently planted, was an early introduction from tropical Asia. This tree, with its enormous fruits sometimes weighing as much as forty pounds, armed with short stout conical protuberances, and often exceeding two to two and one-half feet in length, is an excellent illustration of a common tropical phenomenon known as cauliflory, for the female flowers and fruits are borne directly on the tree trunks and the larger branches. In this species the male flower masses are borne near the tips of the branchlets. The shrubs or small trees known as *carambóla* or *blímbing* (*Averrhoa carambola*, fig. 203B), with its smooth, very acid fruits star-shaped in cross section and borne on the smaller branchlets, and the equally striking *Averrhoa bilimbi*, fig. 203A, with its even more acid, smooth, green, cylindric fruits resembling small cucumbers and borne on the trunks and larger branches (another case of cauliflory), are both Asiatic in origin, but are both widely planted in the archipelago.

While perhaps the majority of the cultivated fruit trees of the archipelago represent early or more modern introductions from other tropical countries, this region also originated a number of distinctly interesting and important species. One of the best of these is the famous mangosteen (*Garcinia mangostana*, fig. 211), strictly Malaysian and ultra-tropical in its requirements; many consider its globose, dark purple fruits, with very delicate, juicy white pulp, to be the very best of tropical fruits, superior even to the mango, to which, however, it is not at all allied botanically. Two other strictly Malaysian fruits are the beautiful red ones of the *rambútan* (*Nephelium lappaceum*, fig. 208A), beset with soft, rather slender, spinelike processes, and almost equally excellent, *pulásan* or *kapulásan* (*Nephelium mutabile*, fig. 208B), its red fruits covered with short blunt tubercles. Both are botanically allied to the equally famous Chinese fruit known as lychee (*Litchi chinensis*). Still another excellent and rather strictly Malaysian native fruit tree, now more or less cultivated outside of the archipelago, like others originating there, is the *lansóne, langsát,* or *dúku* (*Lansium domesticum,* fig. 204). Its pale yellow, ovoid-oblong or subglobose-ovoid, very well-flavored fruits are borne in clusters on the larger branches and even on the tree trunks themselves, another illustration of the tropical phenomenon of cauliflory.

Less desirable, yet commonly cultivated secondary fruit trees are the *santól* or *kechápi* (*Sandoricum koetjape*), with globose, dirty yellow, very shortly pubescent fruits, and several species of *Syzygium* (*Eugenia*), such as the *jámbu áyer* (*Syzygium aqueum,* fig. 217), the *jámbu semárang* or *makópa* (*Syzygium samarangense*), the Malay apple or *jámbu bol* (*Syzygium malaccense*), and the rose-apple or *jámbu áyer* (*Syzygium jambos,* fig. 214), all with pink to red smooth fruits, and the *dúhat* or *jambolán* with oblong-ovoid, smooth, purplish black fruits (*Syzygium cumini,* fig. 215). One of the strange wild but sometimes cultivated trees is the *pángui* or *kepáyang* (*Pangium edule*), characterized by one author as one of nature's monsters. The massive brownish fruit is up to a foot long, with the distinctly large, obscurely triangular, sculptured seeds embedded in a scanty pulp; the seeds are very poisonous when fresh because of the presence of hydrocyanic acid, and yet at times and in certain regions these seeds when dried, washed, and cooked are extensively used as food.

In its way the most famous of the strictly Malaysian fruits is the *durián* (*Durio zibethinus,* fig. 210). This is a tall tree, with large subglobose fruits up to a foot long covered with very hard, sharp, conical processes. This fruit is highly prized by the natives, but its use by Europeans is an acquired taste, for it has been described as having an odor composed of that of many other tropical fruits plus onions, garlic, Limburger cheese, and the smell of the beach at low tide. Once an individual has actually experienced the overpowering odor of this fruit he will never forget it, for when the fruits are ripe and on sale they need no other advertising than their odor, which permeates the atmosphere of the market-place.

In the same genus as the jakfruit mentioned above are various other species of *Artocarpus,* including the breadfruit with its large, prominently lobed leaves and fairly large, globose fruits covered with short or fairly long protuberances. Both the seedless and the seeded forms occur, but as a food the breadfruit is of much less importance in the Malay Archipelago than it is in Polynesia. It is used not as a fruit to be eaten out of hand, but rather as a vegetable—boiled, fried, or roasted. The *champédak* (*Artocarpus champeden*), more closely allied to the jakfruit than to the breadfruit, and like the former bearing its large oblong-cylindric fruits on the tree trunk and larger branches, is also cultivated. It may be distinguished from the jakfruit by its distinctly hairy leaves. Several of the wild species of this genus also have edible fruits or at least edible seeds; the most important of these is the *tempúni* (*Artocarpus rotunda*), with globose medium-sized fruits thickly covered with rigid spines one-third of an inch long in Malaysia, and the Philippine *márang, A. odoratissima,* the latter allied to the breadfruit but with less evidently lobed leaves.

This consideration of trees actually grown for their edible fruits in Malaysia is by no means complete, but the most important and more common species are included. Certain species of *Flacourtia* are sometimes cultivated for their excellent small, globose, smooth, purple fruits, but they more commonly occur wild in various parts of the archipelago, and include the *lovilovi* (*Flacourtia inermis*), the *rúkam* (*Flacourtia rukam*), both trees, and the shrubby and usually spiny *bitangól* (*Flacourtia indica*). Some species of the euphorbiaceous genus *Baccaurea,* known as *rámbai* or *ménteng,* are also cultivated, the fruiting racemes being borne on the branches below the leaves (*B. racemosa*), as well as the peculiar *nam-nam* (*Cynometra cauliflora,* fig. 87) with its strikingly paired leaflets and its yellowish green, acid, inequilateral fruits borne on the trunks and large branches. This peculiar tree belongs in the bean family. Also in the same family is the so-called Polynesian chestnut or *ifi,* which is not a chestnut at all but is rather a bean widely utilized for food in the isuands of the Pacific, with solitary large seeds which are boiled or roasted. This is *Inocarpus fagiferus,* fig. 202, and it is more usually found near the seashore.

ORNAMENTAL TREES

The number of cultivated ornamental plants, trees, shrubs, vines, and even herbs, to say nothing of a great variety of palms and bamboos, is very large. Yet it is of interest to note that the majority of the shade trees and strictly ornamental plants are not native species, as one might expect, but are mostly exotics, originally natives of tropical Asia, Africa, and America. Since it is not possible to consider these *in extenso,* only a few of the more striking ones are mentioned. In the older towns and cities, one of the most outstanding plants that will be observed is the striking traveler's-tree (*Ravenala madagascariensis*) of the banana family; its great banana-like leaves are arranged in one plane at the summit of the trunk, suggesting a great fan in appearance. In most towns one will note the flamboyant, fire-tree, or flame-tree (*Delonix regia,* fig. 234), of the bean family, with its very finely divided compound leaves, long hanging flat pods, and great masses of very large, red or red and yellow flowers; this is a native of Madagascar, introduced into cultivation about a century ago. The African tulip-tree with its very large, irregular, red flowers about four inches long, and large leaflets, is *Spathodea campanulata,* fig. 250. The widespreading shade tree with compound leaves that droop in cloudy and rainy weather, with rather small pink flowers, is the tropical American rain-tree (*Samanea saman*). A smaller tree with much thickened terminal branches rich in milky sap, with elongated entire leaves that often fall in the dry season, and rather large, very fragrant, white and yellow flowers, sometimes tinged with pink, is the frangipani, temple-tree, or *kalachúchi* (*Plumeria acuminata,* fig. 245), a native of Mexico. The *kenánga* or *ílangílang* (*Cananga odorata,* fig. 228) is a native tree, commonly planted because of its dangling, elongated, pale green to dull yellow, very fragrant flowers. The Indian-almond, *ketápang* or *talísay* (*Terminalia catappa,* fig. 42), also a common native of the seacoast, is frequently planted as a shade tree. It is characterized by its rather large obovate leaves and its

THE CULTIVATED PLANTS 157

somewhat compressed and laterally two-keeled, red, one-seeded fruits which are one and one-half to two inches long.

In some regions the *angsána* or *nárra* (*Pterocarpus indicus*), a tree yielding a very fine cabinet timber, characterized by its yellow flowers and its nearly circular, compressed, one-seeded pods surrounded by a distinct wing, is a favorite shade tree, but in many regions it is not so used. A small tree, *katúray* (*Sesbania grandiflora*, fig. 230), with its pure white or pink-red, very large irregular flowers three to three and one-half inches long, and slender hanging pods up to one and one-half feet long, is common. Several species of *Cassia* are not uncommon, the most striking being the pink-shower (*Cassia nodosa* and *C. javanica*) with conspicuous pink flowers, and the golden-shower, *cana fistula* (*Cassia fistula*, fig. 231) with its hanging racemes of yellow flowers. These have very characteristic elongated pendent cylindric pods one to two feet long and three-fourths to one inch in diameter. In this same genus is the even more common *Cassia siamea*, fig. 232, with yellow flowers and flat pods. *Bátai* (*Peltophorum pterocarpum*) is a native seacoast tree frequently planted because of its numerous yellow flowers in erect panicles, its pods being two to four inches long, narrowly but distinctly winged. The tamarind or *sampálok* or *tángkal-asem* (*Tamarindus indica*, fig. 199) will be recognized by its thickened brown pods three to six inches long; the pulp surrounding the seeds is acid. The coral bead-tree or *sága* (*Adenanthera pavonina*) has hard shining bright red seeds. *Lánguil* (*Albizzia lebbek*), not particularly attractive but widely planted, sometimes known as the woman's-tongue tree because of the rather constant clattering of its dry pods which remain on the tree for a considerable period after ripening, will be recognized by its straw-colored, flat pods five to twelve inches long and up to two inches wide, marked with round lumps opposite the seeds.

Champáka or *chempáka* (*Michelia alba*, and *Michelia champaca*, fig. 225) are the cultivated trees that replace our magnolias in the tropics; both have very fragrant flowers, the former white, the latter yellowish or orange-yellow. Kapok or *bóboy*

(*Ceiba pentandra*) deserves mention because of its widely spaced tiers of horizontal branches; the silky fibers surrounding the seeds form the kapok of commerce. Another small tree is the horse-radish-tree, so named because of the taste of its roots. The leaves are compound, the flowers creamy white, and the eight-to fifteen-inch-long fruits hang downward, being triangular in cross section at maturity; this is *Moringa oleifera,* widely known as *malúngai.* Various species of wild figs are planted as shade trees, all in the very large genus *Ficus.* The India rubber-tree (*F. elastica*) with large leaves, the *waringin* (*F. benjamina*) with small leaves, and the sacred banyan or *pípal* (*F. religiosa*), its fairly large leaves with greatly elongated drip tips, are those most commonly used. Many of the wild figs of the banyan type often start as epiphytes and eventually strangle their host tree; they are known as the strangling figs. Others produce their abundant fruits on the trunks and larger branches. In general, the wild figs are characterized by having an abundant milky latex, be they shrubs, vines, or giant trees.

ORNAMENTAL SHRUBS

A considerable number of ornamental shrubs are cultivated, some for their variegated foliage, others for their attractive flowers. Some of these are natives of the archipelago; others were introduced. *Acalypha hispida,* fig. 236, is a commonly planted shrub, characterized by its stout, pendulous, purple spikes. In the group with variegated leaves and often planted for hedges or in fence rows, *Acalypha wilkesiana,* fig. 235, is conspicuous because of its large heart-shaped leaves variously mottled with red, purple, and other colors; other species of this genus in cultivation have green leaves with white margins. It was originally described from Fijian specimens, and its specific name commemorates the commander of the Wilkes United States Exploring Expedition. *Sarása* or *balásbas* (*Graptophyllum pictum,* fig. 252) and several forms of *Pseuderanthemum,* with purple or liver-colored leaves, and in both groups other forms with the green leaves variously mottled white and gray, represent the Acanthaceæ. *Papúa* (*Nothopanax*) with variable leaves, some forms with white-variegated green leaves, is the most commonly cultivated representative of the Araliaceæ.

Everywhere will be noted in cultivation numerous forms of *saguilála* (*Codiæum variegatum,* fig. 237) known to gardeners as "croton"; the leaves range from very narrow to very broad, sometimes twisted like a corkscrew, sometimes with spoon-like tips, and all variously mottled with shades of yellow, purple, and red. This is a species remarkable for variations in leaf form and color. A very commonly planted representative of the lily family is the erect, simple or sparingly branched *handjúwang* (*Cordyline fruticosa,* fig. 220), with its usually purplish, parallel-veined leaves. The bowstring-hemp (*Sansevieria roxburghiana*), with its erect, elongated, fleshy mottled leaves is often common; its fibers are very strong.

Valued for their flowers are various erect shrubs, such as several species of *Ixora,* all with more or less crowded, slender, elongated flowers, some pure white, some pink, some yellowish, and others bright red; a widely used name for these is *síantan* or *sántan*. With attractive red or red and yellow or strictly yellow flowers is *caballéro* (*Cæsalpinia pulcherrima,* fig. 229), a somewhat spiny shrub and one of the earliest introductions into the Orient from Mexico. *Barleria cristata* with its blue flowers, *Tecoma stans,* fig. 249, with irregular yellow flowers, and the narrow-leaved *Thevetia peruviana,* fig. 246, with its bell-shaped yellow flowers—the last two from tropical America—are all common. The white-flowered, very fragrant jasmines include *sampaguíta* (*Jasminum sambac,* fig. 242), usually with double flowers, and *Jasminum multiflorum* with simple flowers. *Kamúning* (*Murraya paniculata*) with its small, white, very fragrant flowers and small red fruits, is also prized for its very hard wood. *Mapóla* (*Hibiscus mutabilis,* fig. 239), six to twelve feet high, a representative of the mallow family, with very large, usually double flowers, opening white or pink and darkening in age, is usually common.

Everywhere one notes the bright red flowers in both single and double forms of the *gomaméla* (*Hibiscus rosa-sinensis,* fig. 241), and its African ally with finely divided petals, (*H. schizopetalus,* fig. 240.). The annatto or *achióte* (*Bixa orellana,* fig. 233) is conspicuous for its reddish brown fruits covered by rather soft spines, the numerous seeds surrounded by thin red pulp much used in coloring food; this is of American origin. *Dama de noche* (*Cestrum nocturnum*), another American contribution to Oriental gardens, has slender, pale greenish flowers which are very fragrant at night but odorless in the daytime. A form of the oleander, with pink and usually double flowers, *Nerium indicum,* is frequently planted. Here may be listed the rather small *Plumbago auriculata,* fig. 243, with its pale blue flowers, and the coarser form, *P. indica,* with its much longer crimson flowers. The African *Thunbergia erecta,* a small shrub with rather slender, elongated blue flowers, is not uncommon in the larger settlements.

Because of their frequent occurrence, not because of any attractive qualities, may be mentioned the croton-oil-plant, *túba* (*Croton tiglium*), the fruits of which, when crushed, are used for stupefying fish, and the very common physic-nut (*Jatropha curcas,* fig. 84) commonly planted in fence rows; but the fairly large seeds of the latter are to be left strictly alone, for, if eaten, they are violently purgative and distinctly poisonous. Various species of *Bauhinia,* some with white flowers (*B. acuminata,* fig. 227), some with lemon-yellow flowers (*B. tomentosa,* fig. 226), and others with purple or pink-purple flowers, are notable (*B. monandra, B. purpurea*), for all of them have characteristic two-lobed leaves. The Mexican *Euphorbia pulcherrima,* fig. 238, a familiar plant at home at Christmastime, is widely planted in most tropical countries, being characterized by its abundant latex and the large, leaf-like, brilliant red bracts subtending the small flowers; this is the *Poinsettia* of our florists. One very curious plant with wide flat green branches, small leaves, or sometimes almost leafless, and small insignificant flowers is *Muehlenbeckia platyclada,* fig. 224.

ORNAMENTAL VINES

In addition to the various trees, shrubs, and herbs cultivated for ornamental purposes are a considerable number of vines with colorful flowers. Among the most striking of these are forms of the Brazilian *Bougainvillea spectabilis,* fig. 223; the masses of colored bracts associated with the flowers persist for a long time after the flowers mature, the colors varying from shades of magenta or lavender-purple to red or even white. More delicate than this is the very attractive *cadena de amor* (*Antigonon leptopus,* fig. 222), a native of Mexico, with masses of pink or white flowers. *Petræa volubilis,* with its attractive racemes of lilac-colored flowers and rather harsh leaves, is not uncommon; this also originated in America. With their lush foliage and large pale blue flowers, *Thunbergia grandiflora,* fig. 251, and *T. laurifolia,* both natives of India, are conspicuous. The tropical American *Allamanda cathartica,* fig. 244, with its attractive large yellow flowers and abundant latex, is frequent. In the morning-glory family, among others, the coarse vine, *Argyreia nervosa,* fig. 247, is conspicuous because of its rank growth, large pubescent leaves, and pink flowers. A strikingly beautiful smaller vine is the African *Clerodendron thomsonæ,* fig. 248, with its angular calyx, white sepals, and spreading dark red petals.

BAMBOOS

Conspicuous in the landscape will be the bamboos. These giant grasses belonging in such genera as *Bambusa, Dendrocalamus, Gigantochloa, Schizostachyum,* and others, sometimes wild, sometimes planted, are in general utilized more for economic than for ornamental purposes. The thin-walled, small or medium-sized species of *Schizostachyum* may at times be gregarious over large areas, and are seldom or never planted. Contrasted to this gregarious type are the larger species which may form large clumps; the basal parts of certain species of *Bambusa* are often supplied with numerous basal spiny branches which form dense thickets and thus more or less protect the clumps against the depredations of animals. The young tender shoots of most of the bamboos are edible.

PALMS

Even more striking than the bamboos will be the palms. They vary greatly in size and appearance, some forming dense clumps, others standing alone. Some of the species are very small in size, while others are actually gigantic. Many are cultivated for ornamental purposes, and others, like the coconut palm (*Cocos nucifera*), the African oil palm (*Elæis guineensis*), and the betel nut palm, *pinang, jambé,* or *búnga* (*Areca catechu*), are grown for their economically useful fruits.

Here the wild palms, in general, are not considered, for scores of genera and several hundred species are involved, but rather this brief treatment is confined largely to planted species, or those that occur, whether planted or self-sown, in the vicinity of settlements. However, everywhere in the forested regions of Malaysia and extending eastward at least as far as Fiji, very numerous species of the striking scandent rattan palms (*Calamus* and *Dæmonorops*) will be observed. Their long stems vary greatly in size, from the thickness of a pencil to a diameter of about two inches. Some of these smooth stems are several hundred feet in length, of the same diameter throughout their length except for a slight thickening at the base and perhaps near the growing tips. The younger parts of these climbing palms are usually supplied with numerous short to long, very stiff, sharp spines. The leafstalks are also spiny, and the lower surface of the leaf rachis is supplied with very stout, sharp, reflexed claws. What will attract even more attention, and will often cause rather violent remarks, will be the long whip-like extensions of the leaf rachis, and sometimes similar organs on the spiny inflorescences; these spiny-clawed, slender extensions are frequently several meters in length and hang down through the forest and along trails wherever these palms occur. They are beset with stout recurved sharp claws which catch one's clothing, and, unless one be wary, produce painful scratches; whenever they take hold of the clothing or skin, discretion is the better part of valor, and it is well to "wait a bit" until one disentangles these exasperating hold-fasts.

THE CULTIVATED PLANTS 163

Palms are always recognizable as such because of their characteristic habit, small to large, unbranched columnar trunks, crowned with their fan-shaped (*búri* or *Corypha*), pinnate (coconut palm), or bipinnate (fish-tail palms or *Caryota*) leaves, with inflorescences and fruits produced either at the tips of the mature trunks, or more generally in the leaf axils, or on the trunk immediately below the green sheathing leaf bases which, in many species, form the characteristic "boot."

PALMS WITH FAN-SHAPED LEAVES

Among the common species with very large fan-like leaves will be the gigantic *Corypha elata,* known as *búri, gébang,* and *ágel,* with its very large cylindric trunks and enormous leaves, and also the almost equally large Palmyra palm (*Borassus flabellifer*), widely known as *lóntar,* and the more slender species of *Livistona,* known as *sádang, sérdang,* and *ánahao.*

PALMS WITH BIPINNATE LEAVES

The only genus with bipinnate leaves will be *Caryota;* some of the species have solitary trunks and others grow in clumps. The palms in this genus are strikingly characterized by their very inequilateral leaflets, which are jaggedly toothed along one of the diagonal margins, whence the common English name of fish-tail palm; the inflorescences are axillary and in the axils of fallen leaves. All of the species with solitary stems die once the cycle of flowering and fruiting is concluded. The fruits contain numerous stinging crystals of oxalate of lime and no attempt should be made to eat them. Widely used local names are *gendúru, sárai, suwángkung, pugáhan,* and *aníbong.*

PALMS WITH PINNATE LEAVES

The coconut palm (*Cocos nucifera*) is the most widely distributed and the most common of the pinnate-leaved species, but other striking ones will be the sugar palm (*Arenga pinnata*), with its enormously long ascending leaves and great clusters of flowers and fruits hanging from the leaf axils, the sheathing parts of the leafstalks being provided with very long stiff black hairs; widely used common names are *hidiók, káong, kábong, arén*. Very commonly planted is the tall slender betel nut palm, *pinang, jámbe, búnga* (*Areca catechu*), the red fruit of which is widely used with air-slaked lime, and the leaves of the betel pepper (*Piper betle*) as a masticatory; this combination turns the saliva red.

Along the lower reaches of tidal streams and always within the influence of salt or brackish water will be often large areas covered with the characteristic stemless nipa palm (*Nipa fruticans*), with its large globose heads of dark brown fruits borne on erect stalks produced from the root. In western Malaysia the striking *Oncosperma filamentosa*, variously known as *érang, liwung*, and *gendiwung*, occurs in similar habitats, but its medium-sized erect trunks bear many long spreading slender awl-shaped or needle-shaped sharp spines; other species of the same genus, all with spiny trunks, occur in the forests here and there. The one striking palm in fresh-water swamps will be the sago palm (*Metroxylon sagu*), widely known as *ságu, lumbía,* and *rumbía,* the source of commercial sago, and a basic food plant especially in the Southwest Pacific region. Spiny and spineless forms occur.

Various species of *Pinanga*, some with solitary trunks, some forming clumps, their leaf segments sometimes rather broad, are chiefly forest dwellers. In the vicinity of the larger towns and cities, various exotic palms have been introduced and occur in cultivation as ornamental species, such as the royal palm (*Roystonea regia*), of the West Indies, *Jubæa* from South America, *Martinezia* from South America, *Latania* from the Mascarene Islands, the African or Indian date palm (*Phœnix*), and others.

The sweet sap derived from the inflorescences of such palms as the coconut, the nipa (*Nipa fruticans*), sugar palm (*Arenga pinnata*), fish-tail palm (*Caryota*), *búri* (*Corypha*), and some other palms, widely known as *túba,* is extensively used as a mildly intoxicating drink when fermented; from it stronger liquors are distilled. The unfermented sap is also widely used

THE CULTIVATED PLANTS 165

as a source of crude sugar, which is obtained by "boiling down" the sap; the method corresponds to that used in the preparation of maple sugar.

HERBACEOUS PLANTS

In general in the tropics comparatively few annual or perennial herbs are cultivated for their showy flowers. Among these may be noted such species as balsam, *Petunia, Cosmos, Coreopsis, Tagetes* (marigold), *Gaillardia,* and *Portulaca. Rhœo discolor,* fig. 219, a somewhat fleshy plant, its leaves purple beneath and its flowers crowded in compressed axillary bracts, is frequently planted. The ordinary cockscomb (*Celosia cristata*) and various forms of *Amaranthus tricolor* with liver-colored to reddish purple leaves are frequent in cultivation, as well as the much smaller *Alternanthera versicolor* of the same family, its small leaves varying from green to purple or red. In general, many of the commonly cultivated ornamental herbs originated in cooler parts of the world, and they naturally do not thrive in the tropics.

Likewise there will be relatively few species of bulbous plants noted in cultivation in the wet tropics, for most of these came originally from cooler regions or from those parts of the tropics having a relatively scanty rainfall. In the calla lily family, *Caladium bicolor,* with its strikingly colored leaves, the colors varying from red to pink, or the leaves green with blotches of red and white, is often common. Most of the bulb plants that thrive in the tropics belong in the amaryllis family; these strongly resemble the lilies but their flowers have inferior rather than superior ovaries. In addition to the tuberose (*Polianthes tuberosa*), one notes in cultivation certain representatives of the genera *Zephyranthes, Crinum, Pancratium, Hymenocallis, Eurycles, Hippeastrum,* and *Eucharis,* the latter being known as the Amazon-lily.

PLANTS CULTIVATED FOR EXPORT PRODUCTS

In the settled areas where European plantations have been established, while extensive areas are naturally devoted to the cultivation of various food plants, and such others as sugar-cane, coconut, African oil palm, and other well-known species, small to very extensive plantings have been established for special crops. Among these are coffee (*Coffea arabica* and other species), tea (*Camellia sinensis*), sugar-cane (*Saccharum officinarum*), coconut palm (*Cocos nucifera*), oil palm (*Elæis guineensis*), quinine (*Cinchona*), pepper (*Piper nigrum*), cubebs (*Piper cubeba*), cacao (*Theobroma cacao,* fig. 209), clove (*Syzygium aromaticum*), ginger (*Zingiber officinale*), nutmeg (*Myristica fragrans*), cinnamon (*Cinnamomum zeylanicum*), tobacco (*Nicotiana tabacum*), kapok (*Ceiba pentandra*), sisal (*Agave*), cassava or tapioca (*Manihot esculenta,* fig. 205), and various others to meet export demands. In the Philippines large plantations of *abacá* or Manila hemp (*Musa textilis*) will be noted; this is a banana with small inedible fruits. The Pará rubber tree (*Hevea brasiliensis*), the chief source of commercial rubber, has been enormously extended in cultivation in Malaysia within the present century.

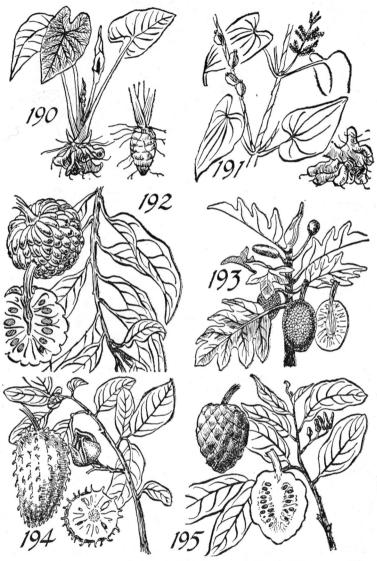

Edible Plants: Fig. 190. *Colocasia esculenta*; Fig. 191. *Dioscorea alata*; Fig. 192. *Annona squamosa*; Fig. 193. *Artocarpus altilis*; Fig. 194. *Annona muricata*; Fig. 195. *Annona reticulata*.

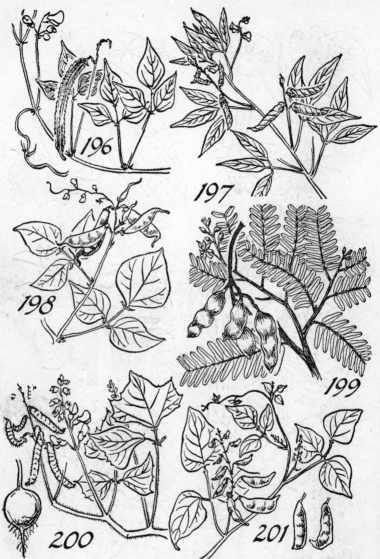

EDIBLE PLANTS: Fig. 196. *Psophocarpus tetragonolobus*; Fig. 197. *Cajanus cajan*; Fig. 198. *Dolichos lablab*; Fig. 199. *Tamarindus indica*; Fig. 200. *Pachyrrhizus erosus*; Fig. 201. *Phaseolus lunatus*.

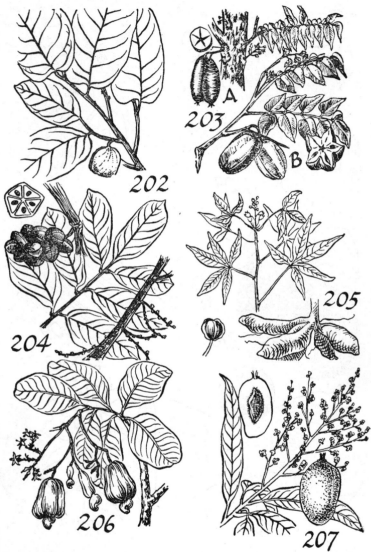

EDIBLE FRUITS, seeds or tubers: Fig. 202. *Inocarpus fagiferus*; Fig. 203. A. *Averrhoa bilimbi*, B. *Averrhoa carambola*; Fig. 204. *Lansium domesticum*; Fig. 205. *Manihot esculenta*; Fig. 206. *Anacardium occidentale*; Fig. 207. *Mangifera indica*.

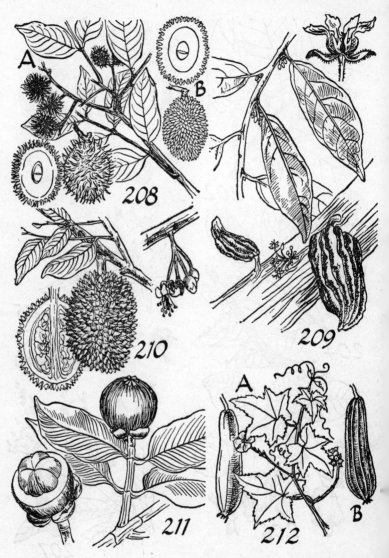

EDIBLE PLANTS: Fig. 208. A. *Nephelium lappaceum*, B. *N. mutabile*; Fig. 209. *Theobroma cacao*; Fig. 210. *Durio zibethinus*; Fig. 211. *Garcinia mangostana*; Fig. 212. A. *Luffa cylindrica*, B. *L. acutangula*.

EDIBLE FRUITS: Fig. 213. *Carica papaya*; Fig. 214. *Syzygium jambos*; Fig. 215. *Syzygium cumini*; Fig. 216. *Psidium guajava*; Fig. 217. *Syzygium aqueum*; Fig. 218. *Achras zapota*.

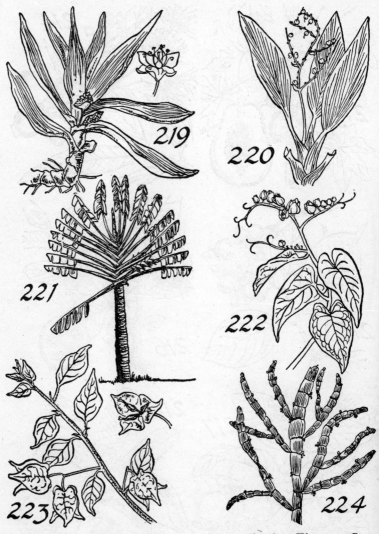

ORNAMENTAL PLANTS: Fig. 219. *Rhoeo discolor*; Fig. 220. *Cordyline fruticosa*; Fig. 221. *Ravenala madagascariensis*; Fig. 222. *Antigonon leptopus*; Fig. 223. *Bougainvillea spectabilis*; Fig. 224. *Muehlenbeckia platyclada*.

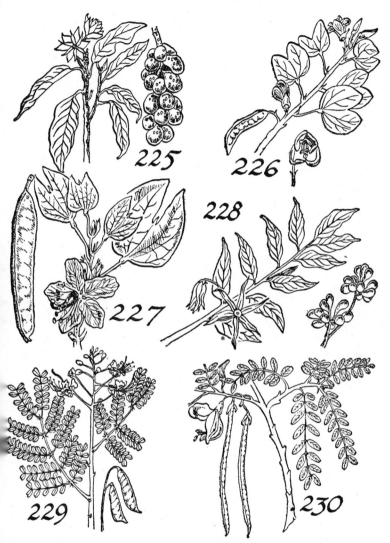

ORNAMENTAL PLANTS: Fig. 225. *Michelia champaca*; Fig. 226. *Bauhinia tomentosa*; Fig. 227. *Bauhinia acuminata*; Fig. 228. *Cananga odorata*; Fig. 229. *Caesalpinia pulcherrima*; Fig. 230. *Sesbania grandiflora*.

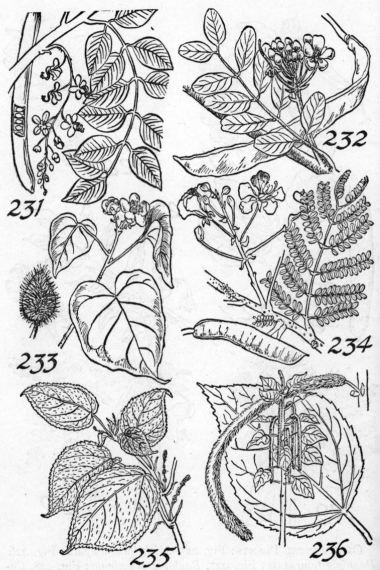

ORNAMENTAL PLANTS: Fig. 231. *Cassia fistula*; Fig. 232. *Cassia siamea*; Fig. 233. *Bixa orellana*; Fig. 234. *Delonix regia*; Fig. 235. *Acalypha wilkesiana*; Fig. 236. *Acalypha hispida*.

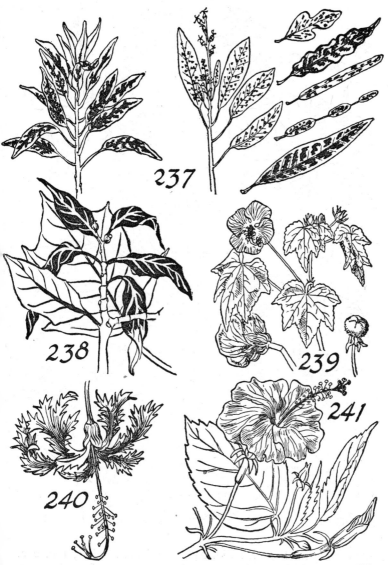

ORNAMENTAL PLANTS: Fig. 237. *Codiaeum variegatum*; Fig. 238. *Euphorbia pulcherrima*; Fig. 239. *Hibiscus mutabilis*; Fig. 240. *Hibiscus schizopetalus*; Fig. 241. *Hibiscus rosa-sinensis*.

Ornamental Plants: Fig. 242. *Jasminum sambac*; Fig. 243. *Plumbago auriculata*; Fig. 244. *Allamanda cathartica*; Fig. 245. *Plumeria acuminata*; Fig. 246. *Thevetia peruviana*; Fig. 247. *Argyreia nervosa*.

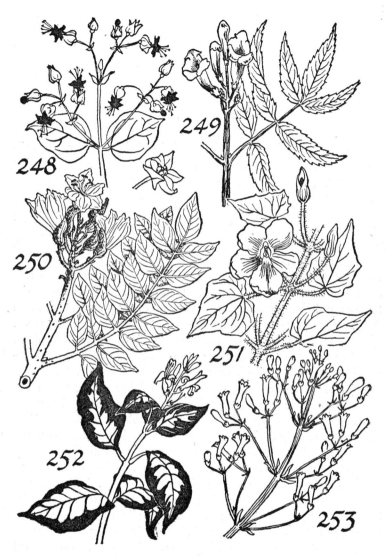

ORNAMENTAL PLANTS: Fig. 248. *Clerodendron thomsonae*; Fig. 249. *Tecoma stans*; Fig. 250. *Spathodea campanulata*; Fig. 251. *Thunbergia grandiflora*; Fig. 252. *Graptophyllum pictum*; Fig. 253. *Russelia juncea*.

10

Jungle Foods

In the tropical forests of the Old World there is always available a considerable number of plants with parts that may safely be eaten, provided one knows what to select and how to prepare it. This applies to various fruits, seeds, underground tubers, the terminal tender buds of various palms (in fact, of most species of palms), the pithy interior of certain palm species such as the sago palm which contains great quantities of starch, the tender shoots of the bamboos, the growing tips and even the roots of some grasses, the seeds of others, the shoots and even the flower buds of the wild and cultivated bananas, the flowers and immature fruits of various other plant species, and the growing parts of a great variety of herbs, shrubs, and trees.

Parts of certain plants, such as the roots of the tapioca plant, certain wild yams, and various seeds, may be distinctly poisonous if eaten when fresh or unprocessed. Other plants, as the vegetative parts and tubers (corms) of the Araceæ and the fruits of certain palms, contain within their tissues innumerable microscopic needle-like stinging crystals of oxalate of lime that are intensely irritating when brought in contact with the mucous membranes. Many of these plant parts are extensively utilized by the natives as food after a simple processing to eliminate the poisonous or irritating principle. This process takes the form of thinly slicing the parts to be used and soaking them for some time in fresh water, or slicing and thoroughly drying and then cooking, or simply cooking for a prolonged time. In soaking or cooking, it is always desirable to use several changes of water. It is always best to seek the advice of natives whenever possible. Even in times of the greatest emergency, where no food seems to be available, the tender twigs of a great variety of plants may be chewed, for something is better than nothing, and such twigs contain some starch and other nutrients. Here one would, of course, avoid those plants that are too bitter or which otherwise have a disagreeable taste.

SOME FOOD AVAILABLE EVEN ON SMALL UNINHABITED ISLANDS

Even on small islands that support very poor floras there are present, on those that have any vegetation, a certain number of species with parts that may be eaten with entire safety. Such are the tender terminal buds of the common screw-pine or *pándan* (*Pandanus,* fig. 11), as well as its fruits and seeds, succulent shore plants such as the seaside purslane (*Sesuvium portulacastrum,* fig. 20), the true purslane (*Portulaca oleracea,* fig. 122), and even, at first such an unpromising plant, the very common *Boerhavia diffusa,* fig. 17, and others. After all, there is no need of starving to death in the midst of relative plenty, even though what one finds to eat may be entirely different from his accustomed diet.

POSSIBILITIES OF JUNGLE FOODS

In this discussion no account is taken of such obvious sources of food as small animals, birds, fish, crustaceans, molluscs, and other forms of animal life. Plants alone are considered, not those that are actually cultivated for food by the natives (see Chapter 9), but rather plants that grow wild along the seashores, in waste lands often as weeds, and in the primary and secondary forests. While it is true that an army could not possibly subsist on what food may be obtained in the jungles, individual men and small bodies of men, separated from their commands, or cast ashore without supplies, can at least maintain life for considerable periods of time.

I am reminded of one episode mentioned by Dr. H. J. Lam in his account of his arduous trip up the Mamberamo River, New Guinea, in 1920. Two Chinese bird hunters accompanied by nineteen native Papuans, a party of twenty-one, traveled overland from Hollandia on Humboldt Bay to the Idenburg River, far in the interior, and thence down the river in proas that they constructed in the forest. According to their own accounts, they subsisted for seventy days on such food as they could secure in the forest and from the river, including small animals, birds, fish, sago from the sago palm, and parts of various other edible plants that occur in the forests. Of course, they had the advantage of knowing what could be eaten with safety, what might be found in reasonable abundance, and where to look for definite species.

AIDS TO CASTAWAYS

Certain unusual aids have been evolved. Thus may be mentioned the Merrill Technical Manual 10–420, *Emergency Food Plants and Poisonous Plants of the Islands of the Pacific,* issued by the War Department and available through the Superintendent of Documents, Government Printing Office, Washington, D. C. This is a pocket booklet of 149 pages with illustrations of 113 plants. Another is the Dahlgren-Stanley *Edible and Poisonous Plants of the Caribbean Region,* 102 pages, with 72 illustrations, published by the Navy Department and available at the same source as the one above mentioned. A smaller useful one is *Edible, Poisonous, and Medicinal Fruits of Central America,* issued in the Panama Canal Zone, with 42 excellent illustrations. Independently of these booklets the Australian government issued *Friendly Fruits and Vegetables,* 71 pages, 37 figures; and in New Zealand *Food is Where You Find it. A Guide to the Emergency Foods of the Western Pacific,* 72 pages, with about 50 plant species illustrated, is available. Naturally, for the most part the same species are included in all of the publications that appertain to the Old World, for the number of potential food plants in the jungles and forests is limited as compared with the total number of plant species that will be found in any one area.

Two excellent compilations including many data regarding both plants and animals available for food, and much additional information of a desirable nature are: *How to Survive on Land*

and Sea. i-xii. 1–264. fig. 1–322. 1943. U. S. Naval Institute, Annapolis, and *Survival on Land and Sea.* i-iii. 1–187. fig. 1–64. 1944. Navy Department, Washington.

SEEK ADVICE FROM NATIVES

Whenever it becomes necessary for anyone to attempt to live off the country, advice from natives regarding potential food plants should be secured if possible. Almost everywhere, particularly where the vegetation is at all well developed, there are few to many species of wild plants, parts of which may be eaten not only with entire safety but often with real pleasure. In those regions where the native population lives fairly close to the earth, a high percentage of the people are thoroughly posted on what may be eaten, and how plant parts should be prepared for food, if the crude product is of such a nature that special preparation be needed to eliminate poisonous or irritating elements. After all there is little need for an individual to starve to death in the midst of relative plenty even although under special circumstances he may be forced to eat types of food that he would not touch at home.

FERNS

The tender unfolding fronds of many different kinds of ferns may be eaten, the so-called croziers or fiddleheads, for these are, in general, non-poisonous. Many are too tough to be eaten, and others may be ill flavored. In Malaysia the very common and widely distributed *páko* or *páku* (*Athyrium esculentum*), frequently abundant along streams and in open wet places, is commonly eaten both raw as a salad and cooked as greens, and is even sold in the native markets. In brackish marshes the very coarse tufted *Acrostichum aureum,* fig. 66A, occurs, sometimes known as *lagólo* and *páku-laut,* and is usually very abundant. Its fronds may be as high as a man, the mature leaves being very leathery; the tender young parts are edible. The same is true of the climbing *diliman, lamidiang,* or *hagnáya* (*Steno-

chlæna palustris, fig. 66B), which also grows in and near brackish marshes. This is a fern sometimes as much as fifteen feet high, with pinnate leaves; the young growing tips and young fronds may be eaten. Even the young fiddleheads of the characteristic tree ferns of the forested regions may be utilized.

PALMS

In the Old World tropics the coconut palm (*Cocos nucifera*) is about the only one with fruits that may be eaten, for the fruits and seeds of most other palms are too hard. However, the scanty acid pulp of all the climbing rattan palms (*Calamus* and *Dæmonorops*) is perfectly safe. Parenthetically it may be stated that these climbing palms are also an excellent source of safe drinking water. Sections should be cut about ten feet long; from the lower end of these, when held in a vertical position, clear water will trickle for a time and then cease. When the flow stops, lower the section, cut off the upper foot or so, replace in a vertical position, and the trickle will again commence, for capillary action must be overcome. Repeat the process as necessary.

Although the fruits of some species, notably the fish-tail palms (*Caryota*), known as *aníbong, pugáhan,* and *gendúru,* are sometimes succulent, they should never be tasted, for the pulp is charged with myriads of minute stinging crystals that are intensely irritating when brought in contact with mucous membranes. The same is true of the even more irritating fruits of the sugar palm, (*Arenga pinnata*), known as *kaóng, ídiok,* and *áren,* but these globose fruits and their included seeds are so hard that it is not likely anyone would even attempt to eat them. Generally speaking, the seeds of all the wild palms are so hard and flint-like that they cannot possibly be eaten. An exception here is the nipa palm (*Nipa fruticans*) that occurs in great numbers in brackish swamps and along tidal streams, for the fairly large *immature* seeds of this palm may be eaten; the mature white seeds are, however, too bony, being much harder than the flesh of a mature coconut palm.

MILLIONAIRE'S SALAD

The *úbud* or palm cabbage is the tender growing part of the palm within the terminal crown of leaves. That of most species of palms is an excellent food, eaten either raw or cooked. In fact, that of the coconut palm is so good that it is sometimes spoken of as "millionaire's salad," for to secure this *úbud* or palm cabbage it is necessary to destroy the palm. Some of these terminal buds, such as those of the coconut palm, the sugar palm, the giant *búri* or *gébang* (*Corypha*), are large in size; others, such as those of the rattan palms, are, of course, small. This potential source of food in the forests should not be overlooked, for palms of one type or another are frequently abundant.

STARCH FROM PALMS

Perhaps the greatest amount of food to be obtained from the palms is in the form of starch. Certain palms, such as the sago (*Metroxylon*) that is frequently very abundant in fresh-water swamps (this is the source of commercial sago), the sugar palm (*Arenga*), the fish-tail palm (*Caryota*), and the giant *búri* or *gébang* (*Corypha*), develop over a period of many years and store up enormous quantities of starch in their trunks. At full maturity these palms produce flowers and fruits and then the plant dies; the stored-up starch is utilized by the plant for flower and fruit production. The interior of these frequently large trunks is more or less pithy, the softer parts being traversed with numerous tough fibers.

The normal procedure in extracting the starch is to fell the mature palm before it commences to produce flowers, crush the softer interior parts, and wash out the starch and allow it to settle in receptacles. The receptacles used are frequently the hollow trunks of old palms, or troughs prepared for the purpose. In times of real emergency, slices of the soft inner parts of these palms may be roasted or boiled and the cooked starch chewed out. Incidentally, the slightly swollen bases of many of the rattan palms contain appreciable quantities of starch, and sections may be roasted and the starch chewed out. All one needs is good teeth.

GRASSES

The grasses, the source of all of our cereals, are safe, although many of the species produce such small seeds that it is rather impracticable to gather them. However, the seeds of Job's-tears (*Coix lachryma-jobi*) are fairly large, but they are difficult to extract from the hard white involucre or receptable inside of which they are produced. Seeds of some of the wild species of *Setaria*, the group that produces Italian millet, are fairly large, and these are, in general, easily gathered. The hearts of the young shoots of the wild sugar-cane (*Saccharum spontaneum*), known as *taláhib kúnai*, or *gelágah*, a very coarse grass six to twelve feet high, with rather harsh leaves, frequently occupying vast areas of open land, may be eaten, as well as its young unopened inflorescences (cooked); and even the underground stems or roots, which are somewhat sweetish, may be peeled and eaten raw. Generally speaking, however, it will be found that the bamboos will yield the most food, for the young tender shoots of these giant grasses may be eaten with entire safety; the young basal shoots develop rapidly and soon become too hard to eat.

AROIDS

This is the group or natural family, the Araceæ, characterized by the calla lily and the Indian turnip (*Arisæma*). In the tropics various species are widely cultivated for food, particularly the *gábi* or taro (*Colocasia*, fig. 190), for its large underground part, the corm, is rich in starch and is the most common substitute for the potato in the tropics. The tropical American yautia, a similar but larger plant of the genus *Xanthosoma*, now occurs here and there in cultivation in the Old World, but is not as yet so widely distributed as the taro. Among the wild species are the giant-leaved species of the elephant's-ear (*Alocasia*), but all representatives of this genus are charged with myriads of very irritating needle-shaped microscopic crystals, like the large corms of the peculiar *Amorphophallus*, fig. 100, known as *kungápung, átjung,* and *tjúmpleng;* yet even the tender young leafstalks of this strange plant may be cooked and eaten, as one would prepare asparagus. The normal procedure in preparing the starchy parts of these plants for food is to slice and dry the parts, wash them thoroughly, and cook them long enough to destroy the stinging crystals. Here, however, it is better to seek the advice of natives before attempting to eat such plants.

OTHER TUBERS

In the forests, various species of wild yams will be observed; sometimes they occur also in thickets and secondary forests. The true yams, not to be confused with the sweet potato (the latter will be found only as a cultivated plant), all belong in the genus *Dioscorea*. Some of these, particularly the *úbi* or greater yam (*Dioscorea alata*, fig. 191), are widely cultivated for food, for their great starchy tubers with the flesh often pale purplish in color, may weigh up to forty or fifty pounds. Most of the tubers of the wild yams are edible and may be eaten with safety after cooking, but others, particularly the *námi* or *kálut* (*Dioscorea hispida*, fig. 68) are distinctly poisonous and may be eaten only after the thinly sliced tubers have been soaked in running water for a considerable period of time, followed by cooking. This treatment eliminates the poisonous principle. Some of these wild yams have developed peculiar adaptations for the protection of their underground parts against wild hogs. In some species the tubers are very deep in the ground—sometimes three or four feet below the surface; others like the *túngo* or *dágo* (*Dioscorea esculenta*, fig. 67), have developed a crown of special very spiny underground stems above the tubers.

The Polynesian arrowroot (*Tacca pinnatifida*, fig. 15) is another strange plant sometimes found in abundance in loose soil in partial shade back of sandy seashores, and occasionally also inland. Its erect green fluted stem bears the large three-parted leafy portion, the sections again variously lobed and divided. The flower stalk grows up from the base of the plant and is as tall as or taller than the leafy part. The hard, often nearly round tubers are borne on short stolons in the loose sand near the base of the plant, sometimes at depths of about a foot. Here again, the fresh tuber should not be eaten, for it is poisonous. Long cooking will dissipate the poison, or treatment may be by slicing, drying, crushing, thorough washing, and then cooking. Here again, if possible, get the advice of natives. The yam bean (*Pachyrrhizus erosus,* fig. 200), a blue-flowered vine with a turnip-shaped root, is a native of tropical America, now more or less widely naturalized in Malaysia. Its crisp roots are eaten raw, and are very refreshing.

TREES AND SHRUBS WITH EDIBLE FRUITS OR SEEDS

A considerable number of trees occur in the forested regions that produce edible fruits and seeds. Among various purely sylvan species will be the tropical representatives of oak (*Quercus* or *Lithocarpus*), always recognizable by their characteristic acorns, and the allied representatives of another genus in the same family very similar to the chestnut; these will belong in the genus *Castanopsis,* and the seeds of some of these are excellent. Some of the tropical acorns are too bitter to be palatable, like those of some of our North American oaks. The large seeds of the so-called Polynesian chestnut (*Inocarpus fagiferus,* fig. 202), a tree having nothing to do with the true chestnut but belonging in the bean family, are excellent when boiled or roasted. This is widely known as *ifi* or *ihi* and occurs chiefly near the sea. The pods bear a single large seed.

A very large tree with compound leaves and large red fruits that split open along one side is *kalúmpang* (*Sterculia fœtida,* fig. 34). Its fairly numerous oblong black seeds are rich in oil, and may be eaten raw or cooked. Even more common than this and strictly a seacoast tree, except where it has been planted inland, is the Indian-almond, *ketápang* or *talísay* (*Terminalia catappa,* fig. 42), with fairly large leaves, slender spikes of small flowers, and red compressed and somewhat keeled fruits usually about an inch long. These fruits contain one excellently flavored seed of small or fair size. The surrounding parts of the fruit are more or less corky, an adaptation to the dissemination of the fruits by floating. Many other species of the same genus occur in the forests, and the seeds of all of them are edible. In the forests also will be noted many species of the genus *Canarium*. The trees all have pinnate leaves and are characterized by the presence of a fragrant resin in the bark. The ovoid fruits vary greatly in size, from one-half inch to perhaps two inches in length; the inner very hard part of the fruit contains a single oily but well-flavored seed. Those species producing large seeds, such as the *kánari* (*Canarium commune*, fig. 70), the *píli* (*C. ovatum*), and others, are highly prized by the natives; thus in the Solomon Islands the seeds of various species of *Canarium*, known as *náli*, are important native foods.

In thickets near the seashore and in some inland forests the palm-like *Cycas circinalis*, fig. 12, may be noted, with its usually unbranched trunks, dense crown of stiff palm-like leaves, and its ovoid one-seeded fruits borne on special inflorescences among the leaves. This is the *fádang* of Guam, and its fruits are commonly used as an emergency food. However, the fresh seeds are very poisonous because of the presence of hydrocyanic acid. The solitary large seeds are crushed, washed thoroughly in several changes of water, dried, and then cooked, to eliminate the poison. *Never eat fádang seeds raw.*

Another source of food in the Malaysian forests is the large-fruited *Pangium edule*, widely known as *pángui*. Its peculiarly sculptured, rather large seeds are not infrequently found along the seashores, but like the seeds of the *Cycas* they are heavily charged with hydrocyanic acid, are hence very poisonous, and *should never be eaten raw*. The treatment to eliminate the poison is similar to that of *Cycas* seeds.

Not uncommon in many forests is the small tree generally known as *bágo* (*Gnetum gnemon*, fig. 69), with its attractive glossy foliage and red, ovoid or ellipsoid, one-seeded fruits. This is a dual-purpose tree, for its seeds may be eaten, boiled or roasted, while the younger leaves are commonly eaten as a substitute for spinach; in fact, it has been designated as "New Guinea cabbage," but it occurs all over Malaysia. Other species of this same genus are large woody vines, but the seeds of all of them are edible. In forests and thickets will be noted various representatives of the genus *Antidesma,* all shrubs or small trees, which produce a profusion of small, red, one-seeded, acid fruits; the rather scanty pulp of these fruits is edible. Even more common than these, especially in forested regions, will be many different species of the myrtle family, representatives of the very large genus *Syzygium* (*Eugenia*), the white to pink or red flowers having many stamens. While the fruits of some of these are perhaps too dry to be used as food, none of them are poisonous, and the pink, red, or dark purple fleshy fruits of some of the species are well flavored; a half-dozen species of this genus are more or less common in cultivation. Two or three species of *Spondias* occur, the so-called Polynesian plum or *vi* (*Spondias dulcis*), and its Malaysian ally (*S. pinnata*), known as *libas;* the rather scanty pulp surrounding the single large seed of the plum-like fruits is edible.

In the forests a considerable number of species of the breadfruit genus (*Artocarpus*) will be noted, and in many places even the wild forms of the breadfruit itself (*A. altilis,* fig. 193). In Polynesia, however, the breadfruit occurs only as a planted tree, and there usually only seedless forms are encountered. The various species of *Artocarpus* are all trees with an abundant milky sap, medium-sized to very large leaves, and usually globose to ovoid fruits. These fruits are often covered with short, blunt processes, which in some forms are elongated and almost spine-like. Some of the species are actually cultivated for their edible fruits, but the large seeds of all of them may be freely eaten either boiled or roasted. They distinctly resemble cooked chestnuts in flavor. In the Malaysian forests (but not in Polynesia) various species of wild mangos (*Mangifera*) are sometimes common. These are large trees, their smooth one-seeded fruits often with a strong turpentine odor. It is a well-known fact that the sap of certain of these species, especially those known as *binjai* (*Mangifera cæsia*), *lánjut* (*M. lageni-*

JUNGLE FOODS

fera), *báchang* (*M. fœtida*), and *kwini* (*M. odorata*), is a contact poison, causing bad skin eruptions (see pp. 4, 5); yet in spite of this, they are actually cultivated by the natives for their edible fruits. One good bit of advice regarding the mangos as a group is to be suspicious of those that have a very strong turpentine odor; and yet the mature fruits of the species that are actually contact poisons may be eaten with entire safety. The large seeds are not eaten.

INTRODUCED AND NATURALIZED FRUIT TREES

Here and there, at least in fairly well-settled areas, but never in the primary forests, certain introduced and naturalized species that bear edible fruits will be noted. The most frequent of these is the American guava (*Psidium guajava*, fig. 216), which is often distinctly abundant, as its seeds are widely scattered by birds and by various animals, and the American papaya (*Carica papaya*, fig. 213), recognizable at once by its normally unbranched, conspicuously scarred, rather soft trunks, terminal crown of large lobed leaves, and yellow or greenish yellow, large, smooth, melon-like fruits borne on the trunk below the leaves. In some places the American cashew (*Anacardium occidentale*, fig. 206) is also naturalized. The large fleshy and juicy basal parts of the fruit are edible and very refreshing, but the seeds, borne in the kidney-shaped terminal part of the fruit, must be boiled or roasted before eating; and as the sap in this fruit wall may cause bad skin eruptions, in boiling or roasting these seed-bearing parts, one should carefully avoid standing in the steam or smoke.

THE PANDAN OR SCREW-PINE

A very widely distributed group is the strange genus *Pandanus*, shrubs or trees with trunks usually supplied with characteristic prop roots, the tips of the branches bearing tufts of long, spiny-margined, spirally arranged leaves. Roughly about 200 different species are known. The usually red or orange-red fruits, made up of a large number of separate parts, vary enormously in size, from an inch or two in diameter to one and one-half feet or more in length; they may be globose, ellipsoid, or cylindric in shape. They grow in the forests and along the seashore; the most common and widely distributed species (*Pandanus tectorius*, fig. 11) occurs often in dense thickets immediately back of sandy seashores, all the way from Madagascar to Hawaii. Thus even on quite uninhabited islands this peculiar plant may be expected, as its fruits are widely scattered by ocean currents. The orange-red pulp surrounding the individual parts of the fruits is well flavored, and is entirely safe to eat. The solitary to rather numerous seeds, often a bit difficult to extract, are well flavored and may be eaten in any quantity; the tender growing tips or "cabbage" deeply buried in the terminal parts of the leafy branches are crisp and tender and may be eaten raw or cooked. Even when no fresh drinking water is available, small quantities of moisture may be had by chewing the rather tender tips of the prop roots. In New Guinea the long cylindric soft red fruits of *Pandanus conoideus* are boiled and eaten by the natives wherever the species occurs, while at middle altitudes other very tall species known as *mondóa, hína, katúra,* and *werámo* (*Pandanus julianetti, P. brosimos*) are extensively planted for their large oily seeds.

POSSIBILITIES OF THE BANANA

Here the banana may be briefly considered. This occurs both as a cultivated plant with many different varieties, and also wild, sometimes in ravines in the primary forests and sometimes in more open country where the forest has been destroyed. It is scarcely necessary to discuss the fruit, for it is well known. In times of emergency the tender parts of the inside of the young basal shoots may be cooked and eaten. Throughout Malaysia the terminal hanging cone-shaped flower buds of certain varieties of banana, known as *póso,* are sold in the native markets and are widely used as a cooked vegetable. The less bitter forms may be compared with our globe artichoke, and are really excellent. In some of the varieties too much tannin is present, resulting in a bitter taste; but cooking in several changes of water will eliminate much of this objectionable feature.

SPINACH SUBSTITUTES

The tender parts, twigs and young leaves, of a very large number of different plants are cooked and eaten by the natives, and some of these are really first class in all respects—as good as or better than spinach, which perhaps isn't saying too much, for many people dislike even the thought of spinach. In this general category may be mentioned again the tender terminal buds of various palms, the so-called palm cabbage, and the corresponding but smaller tender growing tips of the common screw-pine (*Pandanus*). The edible qualities of the younger leaves of the *Gnetum* tree are mentioned above. However, the leaves and tender twigs of other shrubs and trees, such as those of some species of *Acalypha* of the Euphorbiaceæ, *Pisonia* of the Nyctaginaceæ, the horse-radish-tree widely known as *malúngai* (*Moringa oleifera*), the very large flowers of the *katúrai* tree (*Sesbania grandiflora,* fig. 230), the leaves of *bangkúdo* (*Morinda citrifolia,* fig. 49), those of the seacoast *banálo* (*Thespesia populnea,* fig. 38), of the coral-tree or *dap-dap* (*Erythrina variegata,* fig. 27), with conspicuous large bright red flowers, and others may be utilized as food. The small acid leaves of a tree which occurs only along the seashore (*Pemphis acidula,* fig. 36) may be eaten raw.

Among the herbs, a great many different ones are edible when cooked, or for that matter, even raw. Some of these are sylvan; others occur as common weeds and frequently in great abundance in the open country. They include the common purslane (*Portulaca oleracea,* fig. 122), characterized by its thick succulent stems and leaves and its yellow flowers; the so-called seaside purslane (*Sesuvium portulacastrum,* fig. 20), which grows only in close proximity to the sea; various representatives of the Amaranthaceæ, all common weeds, particularly representatives of the genus *Amaranthus,* figs. 118, 120, as well as the very common *Alternanthera,* fig. 112, of the same family; many of the Commelinaceæ, figs. 114, 116, 117, mostly prostrate or spreading herbs with parallel-veined, somewhat fleshy leaves and small blue flowers; the abundant *Boerhavia diffusa,* fig. 17, common along some seashores, a diffusely spreading broad-leaved herb with small pink flowers; *Solanum nigrum,* fig. 169, with its small white flowers and small globose fleshy black fruits often abundant in waste places; and such other common weeds as the pink-flowered *Emilia sonchifolia,* fig. 184, a small plant with scanty milk-sap botanically allied to lettuce, and in this same family (the Compositæ) the introduced and frequently very abundant *Erechtites,* fig. 185, with pink flowers, *Spilanthes acmella,* with yellow flowers, and others. Here again, the natives are usually well posted as to what can be eaten and how the material should be prepared.

Among the commonly cultivated plants, the tender growing tips of the sweet-potato vine and similar parts of the common squash vine are universally used as greens or pot-herbs, supplementing various species that are grown only for this purpose. It is admitted that the nourishment to be derived from such plant parts is not great, yet greens serve their purposes, and among other things provide certain essential vitamins.

11

Problems of Malaysian Plant Distribution

CENTERS OF ORIGIN

Those who are reasonably familiar with the vegetation of the very large Malay Archipelago realize that it is one of the great world centers of origin for both genera and species of plants. Within the region itself, two sub-areas seem to be clearly distinguishable, one centered around the Sunda Islands —Sumatra, Java, Borneo, and of course the Malay Pensinula— and the other about New Guinea and its adjacent islands. Within this area it is estimated that at least 45,000 different species of the higher plants, including the flowering plants and the ferns and fern allies, occur. While the climatic conditions are strictly tropical, except at higher altitudes where temperate conditions obtain, there are vast differences in the amount and seasonal distribution of the rainfall; some restricted areas, like parts of the Lesser Sunda Islands and Timor and certain parts of New Guinea have much less rainfall than other parts of the region as a whole. Again, and this factor has a very decided effect on the distribution of vegetative types, vast areas are subject to alternating and prolonged wet and dry seasons, and yet nearly everywhere in the archipelago are greater or smaller areas where the precipitation is more or less uniform through all parts of the year; and naturally, there are all degrees of transition as between the constantly wet, and the alternating wet and dry conditions. Then there is to be taken into consideration the variations in the soil, for some plant species are adapted to acid soils, others to neutral or alkaline ones, some to limestone formations, others to light sandy soils. This combination of factors, including high temperatures, high humidity, abundant precipitation, topography, altitude, variation in soil conditions, and varying degrees of isolation within what is now the archipelago in past geologic time, has favored the development and perpetuation of an exceedingly rich and exuberant flora, and at the same time has fostered the development of a remarkably high local endemism.

STRATEGIC POSITION OF MALAYSIA

As one scans lists of genera and species characteristic of the Old World tropics as a whole, one soon realizes that the Malay Archipelago is located in a most strategic position in reference to the vegetation of tropical Asia to the north and Australia to the south. Clearly a great many elements are common to the archipelago and the Asiatic continent, and, as would be expected from geographic proximity, there are also indications of plant relationships as between the archipelago and Australia; and yet, in general, one is impressed with certain distinct trends in the distribution of representatives of individual genera within the archipelago as a region. Thus many Asiatic-Malaysian groups are strongly represented in the northern and western parts of Malaysia and tend to become less strongly represented as one progresses eastward. Conversely, many typically Australian groups are strongly represented in Papuasia but the distribution of these types within the archipelago as a whole is definitely restricted, very largely to the eastern islands. Nor is it geographic proximity that determines the occurrence of these Australian types in the archipelago, for Java, very much closer to the Australian continent than is the Philippine group, has relatively few Australian types in its flora as compared with the more distant Philippines; and within the Philippines the rather numerous Australian types tend to be fairly well distributed, even to northern Luzon and the small islands between Luzon and Formosa, but they do not reach Formosa.

STRIKING PHYSIOGRAPHIC AND HYDROGRAPHIC FEATURES

A glance at a good relief and hydrographic map of Malaysia reveals the presence of two great continental shelves, one extending to the south and east from Asia, the other extending to the north from Australia. The submarine surfaces of these two continental shelves are very level and even, and the average depth of the water is only about 100 feet. It is a striking fact that a drop in sea-level of only 150 feet would today unite all of the Sunda Islands—Sumatra, Java, Borneo, the neighboring small islands and the Palawan-Calamian group of the Philippines—with southern Asia, and a similar drop of only about 65 feet would unite New Guinea with Australia. In passing, it should be noted that the Asiatic continental shelf, delimiting what has been called Sundaland, is the largest one in the world, extending about 1,500 miles from east to west. The Sahul shelf, delimiting what has been called Papualand, is smaller, yet it is also very striking; both are delimited by the 100-fathom (600 feet) line, the limits of the continental shelves. This intermediate insular area has been designated as Wallacea (fig. 254).

WALLACEA AND ITS SIGNIFICANCE

Lying between these two continental shelf areas is a very characteristic region whose geological history, since the Tertiary at least, seems to have been distinctly different from the two large and apparently more stable areas to the north and west and to the south and east. This covers all of the Philippines (except the Palawan-Calamian group), Celebes, the Moluccas, and the Lesser Sunda Islands as far west as Lombok. This region is characterized by irregular rows of high islands separated by great deeps, the latter sometimes up to 16,000 feet. There are no corresponding deeps in what has been designated as Sundaland and Papualand, for the submarine topography on the continental shelves that delimit these areas is remarkably smooth and level. Apparently all of the islands in this intermediate unstable area have been subjected to great elevations and depressions, not once but several times in past geologic history. Because of this condition, there seem to have been no continuous land connections between Asia and Australia since the close of the Tertiary; and this assumed fact explains why the mammals, in any considerable number, failed to reach Australia. As a group, the mammals are essentially Eurasian and American; their chief development took place too late for many representatives, dependent on land connections, to make the passage to Australia.

However, there apparently were intermittent land connections from one island or island group to neighboring ones within this unstable area at various times. These were apparently more north and south than east and west, for there is a strongly marked route appertaining to many genera, Sunda Islands northeastward into the Philippines, thence south and southeastward to Celebes, the Moluccas, and New Guinea. One curious thing about this assumed major migration route is that the Australian elements that reached the Philippines in distinctly large numbers not only failed to reach the Sunda Islands over a direct east-west route, but also to a very large degree failed to negotiate the route from the Philippines southwestward to the Sunda Islands; some species did, however, reach Borneo. It may be assumed that any land connections that may have existed within the unstable area in Pliocene-Pleistocene times were more or less temporary, and that possible connections at one time favored certain intermigrations, while possible disconnections inhibited others. If this general thesis is correct, then at times certain intermigrations of both plants and animals were favored, and at other times they were inhibited.

LATE GEOLOGIC EPISODES

Regarding the unstable conditions mentioned above, apparently the last great episode in the history of Celebes and Halmaheira (Gilolo) was a depression of both islands, leaving the mountain ridges, the skeletons of both islands, well above the sea, but drowning the valleys which now appear as great reentrant bays. A very slight depression of the island of Mindanao in the Philippines immediately north of Celebes would result in a replica of the two islands just mentioned, for arms of the sea would submerge the great river valleys of the Cotabato in the south and the Agusan in the north. Incident to this discussion, it is well to note the distribution of active volcanoes as of today—north and south through the Philippines, and the assumed unstable area south of the Philippines, then westward through Java and northwestward along the west coast of Sumatra. Active volcanoes are absent within the centers of the assumed stable areas—Borneo in Sundaland and New Guinea (for the most part) in Papualand; but they do characterize the groups of islands associated with eastern New Guinea. This matter of vulcanism is apparently more or less associated with regions where unstable areas impinge on more stable ones.

PLANT AND ANIMAL DISTRIBUTION

Much has been written on the geologic history and the hydrographic features of the archipelago as a whole, and the problems of animal distribution within the region have claimed the attention of many biologists. We may theorize but we cannot absolutely prove any general thesis in reference to the underlying causes that have shaped present-day geography and present-day phytogeography and zoögeography of the area. It is doubtful if any area in the world presents as many fascinating problems in the field of biogeography as Malaysia.

Alfred Russell Wallace, in the ten years that he journeyed as a naturalist through various parts of the archipelago, observed that in the general distribution of Asiatic and Australian types of animal life, the Asiatic types typically extend southward and eastward to Borneo and Bali, and the Australian types tend to extend northward and westward to Celebes and Lombok. On the basis of Wallace's observations, which indicated a demarcation or limitation as between the Asiatic and the Australian influences on the animal life of the archipelago, Huxley proposed Wallace's Line to separate the Asiatic and Australian faunas. As proposed, it passes through the Lombok Passage between Lombok and Bali, thence northward through the Macassar Strait between Borneo and Celebes, and then eastward between Celebes and Mindanao into the Pacific Ocean.

THE WALLACE AND WEBER LINES

Wallace's Line is only one of many that have been proposed to mark the separation of Asiatic and Australian animal life, but all of these were placed in the area somewhat between Borneo and New Guinea. Each proponent of a substitute line was influenced by the geographic distribution of the particular group he was studying—mammals, birds, reptiles, fresh-water fishes, shells, or insects—for these various lines were all proposed by zoölogists. The most acceptable of the proposed substitute lines is that named by Pelseneer as Weber's Line, in honor of Max Weber's observations on problems of geographic distribution of various types of animal life. Weber's Line was placed much farther to the east, and Celebes, the Moluccas, and the Lesser Sunda Islands were left with Asia, while New Guinea, and its neighboring islands, was left with Australia.

MODIFICATION OF WALLACE'S LINE

In earlier times very little was known regarding the Philippine fauna and flora, but in more recent years, on the basis of our very greatly increased knowledge of the Philippines, it has been proposed to modify the northern part of the line, extending it northward through the Sibutu Passage between Borneo and the Sulu Archipelago, thence through the deep Sulu Sea and northward via the Mindoro Strait between the Calamian group and Mindoro, thence still northward along the west coast of Luzon and finally into the Pacific Ocean very close to the south end of Formosa between Formosa and its neighboring small island of Betel Tobago.

NO ABSOLUTE LINE OF DEMARCATION

Now it is very evident that no proposed line of demarcation is absolute when it comes to the consideration of the present-day distribution of all types of animals and plants. This is particularly true of plants, for in general botanists do not trust mere lines of separation, for the simple reason that too many plant species transcend this or that limit. True, the original Wallace's Line is an almost absolute separation, when one considers only the true fresh-water fishes; but neither it, nor Weber's Line, nor any other proposed one, is so nearly absolute

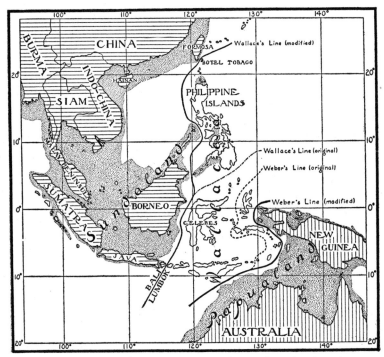

Fig. 254. Chart of the eastern Malaysian region showing the continental shelves, and certain proposed lines that have been designated to separate the Asiatic and the Australian faunas.

for any other group of plants or animals. They at most indicate the trend of limitation in distribution, for in most of the great divisions of the plant and animal kingdom there will be a certain number of types that transcend any proposed line of demarcation. It is significant that the large apes did not extend east of Borneo, in spite of the fact that other large mammals did reach Celebes and other eastern islands; it is significant that certain Australian types of birds did not extend west of Celebes and Lombok. There must be some logical explanation of these distinct trends, for they extend to most major groups in both botany and zoölogy.

SUNDALAND AND PAPUALAND

Now scan a hydrographic and relief map of the entire archipelago that shows the extent of the two great continental shelves above discussed, and impose on this both Wallace's (as originally placed, and as modified) and Weber's Lines. It will at once be noted that in position the former approximates the eastern boundary of the Asiatic continental shelf, in other words, the eastern boundary of the assumed ancient Sundaland; and that Weber's Line approximates the northern and western boundaries of the Sahul shelf, in other words, the corresponding limits of the assumed ancient Papualand (fig. 254). It is reasonably certain that all of the islands on the Asiatic shelf were at times united to the continent of Asia, and at the same time New Guinea and at least some of its neighboring islands were united to Australia. This being so, both plants and animals could then migrate to and from what are now insular areas from the two continental masses over continuous land areas without being inhibited by broad stretches of water, for distribution of both plants and animals over continuous land surfaces is quite a different thing from their migrations from island to island over wider or narrower intervening arms of the sea.

THE UNSTABLE INTERMEDIATE AREA

With this brief review in mind, it is only logical to set up the hypothesis that two former reasonably stable continental areas existed within what is now Malaysia, separated by an unstable area that has had a distinctly different geological history, and one that in recent times and back to the Pliocene-Pleistocene has been constantly insular. This particular area lying between Wallace's Line and Weber's Line has been designated as Wallacea. It includes all of the Philippines (except the Palawan-Calamian group), Celebes, the Moluccas, and the Lesser Sunda Islands, westward to and including Lombok. This intermediate insular area has been populated by descendants of its original fauna and flora plus infiltrations from the west and north and from the east and south; but apparently generalized east-and-west distributions in all groups of plants and animals, from the Sunda Islands eastward to New Guinea and vice versa, have been greatly inhibited since the close of the Tertiary, although not altogether prevented. We may check the known distribution of representatives of various families of flowering plants as well as that of the mammals, birds, reptiles, fresh-water fishes, batrachians, land turtles, land and fresh-water shells, and various groups of insects—and in general, the trend is unmistakably in favor of this idea of an unstable insular area separating two more stable continental areas, rather than a mere line of separation. This idea of an intermediate area agrees best with what we know now of the geological history of the region and the present-day distribution of both plants and animals.

TRENDS OF ASIATIC AND AUSTRALIAN TYPES

Generally speaking, in considering the present known distribution of various types of animals and plants, the Asiatic or continental types reaching the present insular region become progressively less dominant as one proceeds to the south and east, and at the same time the Australian types that are strongly represented in New Guinea likewise become less dominant as one progresses north and west. While there is a more or less universal distribution of representatives of many genera of plants, and various groups of animals, for that matter, from Asia through Malaysia to Australia, there is, however, a surprisingly high number of genera that fail to cover the entire distance. And yet within the archipelago as far as temperature, humidity, amount and seasonal distribution of the rainfall, and special adaptations to the dissemination of plants and other factors are concerned, other than present or past geographic separation, one is at a loss to explain why these numerous genera failed to extend over the entire region; we are forced to seek the explanation of this phenomenon by delving into geologic history.

CONTINENTAL CONDITIONS AND THE NATURAL DISTRIBUTION OF PLANTS

Analyzing our accumulated information still further, it is noted that both low-altitude and high-altitude species are involved, and this applies to both Asiatic and Australian types. One can scarcely escape the conclusion that in past geologic times there have been land connections between what are now strictly insular areas and the continental masses to the north and to the south; and that these ancient connections permitted a certain amount of intermigration of both plants and animals. But these interchanges were in general north and south, for to a certain degree east-and-west migrations, at least in late geologic time, seem to have been inhibited.

We can only assume, *a priori,* that certain genera originated in this or that general region. Within a land mass that has been reasonably stable for a long period of time, we can understand why certain types of vegetation dominate certain parts of a continental area, because of differences in climatic conditions and the presence of mountain barriers or great stretches of arid land, as is the case within the continental United States. If the climatic conditions had been uniform over all parts of North America from the eastern to the western limits of the continent, we would naturally expect to find the same general types of vegetation characterizing all parts of the country, more or less the situation that we do find in northern Canada; for after all, in the north, except for the differences between forest and tundra types of vegetation, there is a remarkably high percentage of universally distributed genera and species extending all the way from Labrador to Alaska. The differences between the Atlantic and the Pacific coastal regions of the United States are very great, for between them lie the great plains and vast semi-arid regions west of the plains.

We can further assume that had continuous continental conditions prevailed in the past, in the region between Asia and Australia, generally speaking, the types of vegetation would have been much more similar from east to west and from north to south than is now actually the case. It is possible that a vast continental area within what is now Malaysia might have included a drier climate here and there, within the limits of any such continent, but this would of necessity be the only factor that could have greatly affected the vegetation, for the whole region is and long has been strictly tropical. There seems to be little reason for considering that such a condition ever existed, except possibly very locally.

CORRELATIONS

In passing, this hypothesis explains reasonably well the presence of very numerous Australian types of plants, both high- and low-altitude forms, far north in the Philippines, as well as certain types of Australian birds and other types of animals, as compared with their great scarcity in Borneo, Java, and Sumatra. By setting up the hypothesis of more or less north-and-south, previously existing connections as opposed to east-and-west ones, it would be possible for Australian types that extended to New Guinea to continue northward through Halmaheira, Celebes, the Sangir Islands to the Philippines; for there is some evidence in the very presence of now submerged ridges extending southward from Mindanao via Sangir to the northeastern tip of Celebes, and from the east-coast range of Mindanao southward to Halmaheira that at least isthmian connections did once exist where now only islands remain. And it is interesting to note that as our knowledge of the flora of New Guinea and its adjacent islands increases, it becomes more and more evident that it has much in common with the Philippine flora, but does not tend to show any striking increase in elements that might be interpreted as of Western Malayan origin.

Theories are always interesting as theories. Sometimes long and universally accepted ones are suddenly overthrown as more information becomes available. Whether or not the ideas expressed are wholly or in part true, they at least seem to be logical in view of what we know at present. In passing, it may be noted that Wegner's theory of drifting continents has been applied to this region, the assumption being that the Australian continental mass moved northward and perhaps westward, and in so doing crumpled the whole southeastern edge of a southern extension of the Asiatic continent into a series of high mountainous islands separated by great deeps, thus permitting a greater or lesser degree of intermingling of the Asiatic and Australian plant and animal life. This is merely an attempt to explain the origin of what we know to exist—an unstable insular area separated by two more stable continental masses, the latter delimited by the existing continental shelves. Like some other theories, its acceptance would explain certain observed phenomena, but at the same time would leave unexplained another great mass of data that does not conform.

12

Problems of Polynesian Plant Distribution

Familiarity with the floras of Malaysia, Papuasia, Micronesia, and Polynesia impresses one very strongly with the evident fact that the vegetation of the Pacific islands is made up, on the whole, of western elements. The Micronesian-Polynesian floras clearly, for the most part, represent the same general types that are characteristic of the Malaysian region, and for all practical purposes the floras of the different islands in the tropical Pacific basin are greatly attenuated Malaysian ones. These floras have largely been derived in ancient times from the larger islands in the west, particularly New Guinea and its adjacent island groups and the Philippines. At the same time, Polynesia and Micronesia, like Papuasia and the Philippines, contain some definitely Australian types, and in the east, particularly Hawaii, some plants that can only be considered as derived from American forms. But even in Hawaii the great majority of the indigenous genera are definitely Malaysian, although most of the species in these genera are different from those that occur in the western Pacific region. At the same time, most of the endemic Hawaiian genera are allied to those that occur in the west.

It has also been suggested that scattered throughout this vast region are certain types that were apparently derived from ancient Antarctica. But this idea is purely theoretical and is one that can scarcely be proved. It is, however, manifest from the extensive coal deposits known to occur in Antarctica that in earlier geological times that region must have supported a very rich flora. If this idea of some Antarctica origins be true for Polynesia, it is also true for the entire Malaysian region. The present restricted distribution of certain characteristic natural groups (families) of plants, to southern South America, New

Zealand, Tasmania, Australia, with extensions northward into the Philippines and Malaysia, and even to the Asiatic continent in some cases, suggests that Antarctica at one time did play an important role in plant distribution.

While there is more or less general agreement as to the origins and relationships of the Polynesian-Micronesian floras, there is no unanimity of opinion as to how numerous genera, many with no manifest adaptations for wide dissemination of their fruits or seeds, reached these now very widely separated islands or island groups. By whatever means the ancestors of the present indigenous plant species reached these distant lands, we can only conclude that it was through natural agencies of one type or another.

THE HIGH ISLANDS

There are several types of islands in the wider reaches of the Pacific, which may be classified in two basic categories—high and low, the names themselves being descriptive. The high islands are relatively few, are wholly or in part of volcanic origin, and as their name implies are characterized by the presence of high mountains or hills; their topography is often distinctly rugged. The chief groups of larger high islands are Fiji, Samoa, Hawaii, Tahiti, and the Marquesas, with smaller high islands in the Caroline and Marianas groups, and such more widely scattered ones as Rapa, Tonga, Rarotonga, Easter, Pitcairn, and others. In contrast with the typical low coral islands, these high ones are in general forested and they support relatively rich floras, as opposed to the poor floras of the low islands. But when one contrasts the flora of any high island in the Pacific basin with that of the larger islands of the Malaysian-Papuasian regions, the contrast is very great; for these remote high islands of the Pacific, without exception, have floras much poorer in genera and in species than those of the larger islands to the west.

THE LOW ISLANDS

The low islands are exactly what their name implies. They are very numerous and are generally associated with atolls or with the remnants of atolls; their presence is due to the growth of coral reefs. They are usually raised only a slight distance above sea-level, and topographically their surfaces are mostly flat or nearly so. Modified low islands are those where the entire island, originally a typical coral atoll or a low coral island, has been raised well above the sea, with resulting modifications. Such islands may be of considerable size and raised a hundred or more feet above sea-level. This type of raised coral limestone island in Pacific Ocean parlance is known as *makatea*, and Henderson Island is typical of this subcategory.

PAUCITY OF SPECIES ON LOW ISLANDS

Botanically, the low islands are very uninteresting and monotonous. The flora of one is usually quite the same as that of another, although these islands and islets may be separated by many hundred and in some cases several thousand miles. The native vegetation may be scanty or reasonably well developed, depending on the size of the island, the quality of its soil, and whether or not it is permanently inhabited.

Just how poor the total flora of an isolated group of small islands may be, in spite of the vegetation being actually luxuriant, is illustrated by Palmyra Island. This group is located about 1,000 miles south of Hawaii and consists of fifty-two islets, the largest being forty-six acres in extent. The highest land is only five feet above sea-level. Its flora consists of only fifteen different species, but most of the several species occur in great abundance. The vegetation has not been disturbed by introduced weeds, as is the case in the not distant Fanning and Christmas Islands. In any case, the total native flora is limited very largely or entirely to those relatively few species characteristic of the strand vegetation of the western Pacific islands (see Chapter 3) minus, of course, the mangrove elements discussed in Chapter 4. It should be noted that not all of the Malaysian strand plants whose seeds or fruits are adapted to dissemination by ocean currents reached the remote small islands in the Pacific, for there are apparently limitations on the time that certain seeds or fruits will float and retain viability.

Among the low islands occasionally one will be found that is well covered by trees of considerable size. The tree in this case is usually a species of *Pisonia* of the Nyctaginaceæ, for its slender fruits are covered with a very sticky substance that causes them to adhere very firmly to the feathers of certain birds, such as the booby, which often nests in the *Pisonia* trees. This is an interesting adaptation to distribution of seeds, for birds are a factor not to be overlooked in the dissemination of certain types of plants—more especially, of course, those plants with seeds transmitted here and there by fruit-eating types. Supplementing the scanty strand species and those distributed by birds, characteristic of the beach and coral sand habitats, will be found various introduced weeds, at least on the inhabited islands, and on these occur also some cultivated plants. The vegetation is normally not dense, and no serious problems exist in reference to getting about on the land.

RAISED CORAL ISLANDS

The *makatea* type is very different from the typical low islands. Frequently these islands are more or less inaccessible by reason of cliffs, sometimes definitely undercut by wave action, arising directly from the sea or from immediately back of the beach. The coral rock exposed to the action of the elements is frequently worn into very sharp, jagged, almost knife-like ridges. Although the vegetation here again consists of relatively few species, it is apt to be exceedingly dense; this makes it very difficult to penetrate the tangled jungles, and these difficulties are increased because of the nature of the underlying coral rock and the frequent presence of larger or smaller pits or depressions. However, on such islands there are apt to be some endemic plant species, and as compared with the absence of endemics on the typical low islands, the vegetation is well worthy of special attention even on the part of the casual visitor.

Some of the raised coral islands are well populated, such as Niue, situated about 300 miles east of Tonga, 350 miles southeast of Samoa, and 580 miles west of Rarotonga. It is about

13 miles long and 11 miles wide. The average elevation is about 225 feet. The total flora on the basis of a very recent publication * is but 460 species, and it is noteworthy that at least half of these represent introduced weeds and cultivated plants.

VEGETATION OF HIGH ISLANDS

The vegetation on the typical high islands is well developed, particularly within the forested areas, but for these high islands within the Pacific basin as a whole the number of endemic genera is relatively small and most of them have definite relationships with those of Malaysia. At the same time, most of the endemic species are also allied to those of the tropical lands bordering the western Pacific. The native flora of these high islands is made up of the same elements that characterize the vegetation of similar habitats in Malaysia, and as far east as Fiji, Samoa, and Tonga one finds characteristic elements of the mangrove swamp vegetation. The genera represented in the forested regions of the high Pacific islands are, for the most part, the same as those that occur within the forested areas of the larger islands to the west, but their number is relatively small. It seems scarcely desirable to attempt a consideration of these high-island forest types, for within limitations, because of the much smaller number of genera and a very much smaller number of species, these data would be quite similar to those appertaining to the primary and secondary forests of Malaysia.

On these high islands of the Pacific basin, both primary and secondary forest types occur, for the secondary forests here, as in Malaysia, owe their presence largely to the activities of man, who has destroyed the original virgin forest to provide land for agricultural development. As in the larger islands to the west, on some of these, notably Fiji, extensive areas exist that are covered with coarse grasses; these grass areas in some places are provided with scattered trees. These apparently natural grasslands are usually found only in those parts of the high islands where there is a more limited rainfall and a prolonged

* YUNCKER, T. G. *The Flora of Niue Island*. Bishop Museum Bull. 178, pp. 1-126, pl. 1-4, fig. 1-3, 1943.

dry season. The elements in these secondary forests are for all practical purposes the same as those characteristic of the larger islands bordering the western Pacific, and this same statement also applies to the open grasslands in general.

One notes in making comparisons that the high islands of any considerable size, such as Fiji and Samoa, are even more closely allied, from the standpoint of plant geography, with the islands of western Malaysia than are the more distant groups, such as Hawaii, Tahiti, and the Marquesas. The farther to the east one travels, the less development there is in genera, with the possible exception of Hawaii; but here geologic age and the total land areas, together with very much higher altitudes than obtain in any of the other Pacific islands, may have played a part.

VEGETATION OF ISOLATED SMALL HIGH ISLANDS

When the total flora of the smaller, more or less isolated high islands is considered, such as those in the Marianas and Caroline groups, and the islands more or less similar in size located still farther to the east, one is impressed with the relative paucity of individual species as compared with what one finds on islands of a similar size located within limits of the Malay Archipelago. Thus with all of the islands under Japanese mandate, and including a number of high, but at the same time relatively small islands, less than 1,300 different species are known, of which 230 manifestly represent purposely or accidentally introduced ones. This relatively small flora includes representatives of approximately 620 genera in 142 families. It should be noted that these figures are based on the study of extensive modern collections, and may thus be accepted as giving a reasonably complete picture of the extent of the present-day vegetation. While the Japanese-mandated areas cover approximately 3,600,000 square miles, and involve many hundreds of individual islands, the total land area is distinctly small, being less than 2,000 square miles. Specific endemism is relatively high, for approximately 460 species are confined to the islands within the area under consideration. The generic endemism is very low;

about seven endemic genera only are involved for the whole group.

COMPARISONS BETWEEN THE FLORAS OF THESE SMALL ISLANDS WITH THOSE OF LARGER ONES TO THE WEST

It seems to be distinctly worth while to compare these figures covering the known flora of many small islands with a total land area of 2,000 square miles or less, scattered over 3,600,000 square miles of the tropical Pacific, with the nearest large group —the Philippines to the west. In the latter group, with a total land area of approximately 114,400 square miles and the islands in a much more compact group, but still covering some 570,000 square miles of ocean, there are at present known about 1,550 genera and in excess of 8,500 species of flowering plants alone, to which should be added about 1,000 species of ferns and fern allies. And this Philippine flora, in spite of the relatively close proximity of the group to southeastern Asia, Borneo, and Celebes, is highly endemic. Even including the species known to have been introduced, specific endemism approximates 68 per cent; or if only those species be considered that are characteristic of the forested areas, specific endemism approximates 84 per cent. By endemic is meant those species that are not only native of a particular region, but actually confined to it, never having been found elsewhere.

Thus there are some definite relationships between any given flora and the size of the island or island groups involved. In general, the larger the island area, the richer the flora. But isolation is also involved. Thus a small high island like Sibuyan in the Philippines, situated fairly close to neighboring larger islands, actually has a flora richer in genera and in species than that of all of the widely scattered and isolated islands within the Japanese-mandated areas of the tropical Pacific.

The very great richness of the total Philippine flora, as compared with the relatively poor total flora of the Pacific groups above considered, is explained in part by geological history, in part by the very much larger total land area, in part by very much higher altitudes (for certain Philippine mountains are

10,500 feet high, whereas the highest altitude in the Japanese areas is scarcely 3,000 feet), but apparently in large part by their relatively close proximity to the larger land masses that in turn support very richly developed floras. In other words, isolation has not been such an important factor as has been the case in Micronesia. Again, the Philippine flora naturally contains a great many continental or Asiatic types that are lacking in Micronesia, and at the same time a much stronger series of Australian types than is found in the scattered islands of the Pacific.

MISCONCEPTIONS AS TO PLACES OF ORIGIN OF SPECIFIC GROUPS

In considering centers of origin and phytogeographic relationships, the individual may sometimes be misled by the paucity of data available at the time when a theory or suggestion was proposed. When a genus is originally described from a certain island or island group, it is apt to be considered as representing a particular restricted flora. Later, when other species of the same group are discovered and described from new regions, one is inclined to consider the new extension of range as actually representing the flora of the island or island group whence the original species came.

Here are three striking examples. The genus *Vavæa* was described in 1846 on the basis of a single species found in Tonga; in 1858 *Couthovia* was described on the basis of one Hawaiian and one Fijian species; and in 1860 *Dolicholobium* was described from a single Fijian species. Up to the close of the nineteenth century, but two representatives of these three genera had been described from outside the limits of Polynesia. Here, then, were three "Polynesian" genera. This is the picture today, indicating the changes in concept as botanical exploration has been extended and intensified within the present century: *Vavæa* now contains seventeen species, six in the Philippines, four in New Guinea, one in Java, one in Borneo, one in the Caroline Islands, one in northeastern Australia, and three in all of Polynesia. *Couthovia* was recorded from New Guinea in 1888 and ten years later from Celebes, and the Celebesian species was found a few years later in the Philippines. Twenty-six species are now known, fourteen from New Guinea and adjacent islands, two from the Caroline Islands, and one from Celebes and the Philippines, one from New Hebrides, one from New Caledonia, and seven from Fiji and Hawaii. *Dolicholobium* is now represented by twenty-two species, fifteen in New Guinea, the Solomon and Bismarck groups, five in Fiji, one in New Hebrides, and one in the Philippines.

Thus, all three of these characteristic genera are not "Polynesian," as was originally supposed merely because their first species happened to have been described from Polynesian material. They are, rather, essentially Malaysian in a broader sense and more narrowly Papuasian-Philippine types. It is highly probable in the case of these three genera that New Guinea was the center of origin, and that the generic ranges were extended north to the Philippines and eastward into the Pacific, but in the case of one genus (*Vavæa*) slightly to the west (Borneo-Java) and to the south (Australia).

LACK OF COMPLETE INFORMATION

In spite of all that has been accomplished in the field of botanical exploration, we are thus handicapped by our lack of definite knowledge when it comes to a consideration of the known distribution of individual genera and species. We do the best that we can on the basis of what is known, but we still lack sufficiently comprehensive collections of reference material. Vast areas remain in the Southwest Pacific and in Malaysia as a whole, that are either very inadequately explored or have never even been visited by a botanist or a collector. The only extensive areas that can be considered as reasonably well explored from a botanical standpoint are Java, the Malay Peninsula, and most of the Philippines.

LOGICAL AND ILLOGICAL CONCLUSIONS

Much has been written as to whence the Polynesian flora was derived; some explanations are logical, others utterly illogical. Much of the reasoning has been based on too little knowledge of what has been published, and at times it seems to be clear that attempts have been made to fit the known factors to a preconceived theory. In general, when this was the case, distributional and other data that are contrary to the preconceived idea were ignored or overlooked. After all, logical conclusions can be drawn only on the basis of the known geographic distribution of all groups of plants and animals, taken into consideration with what is known of the geologic history of the region. When this is done, we may be able to develop a fairly accurate picture or theory. But every investigator should be careful to avoid wide generalizations on the basis of actual distribution of the representatives of any single natural group.

ON METHODS OF PLANT DISTRIBUTION

All are in agreement that certain types of plants have been widely distributed through the ability of their fruits or seeds to float for long periods of time; this is a natural adaptation for most species characteristic of the strand flora. But this is a very limited flora, consisting of few genera and few species. Many groups of plants are manifestly bird-distributed. Wind has also been a factor, but there are definite limitations here, and few seeds are actually adapted to very wide distribution through the medium of the wind. In modern times there is, of course, man himself, and man at the present time is a very important factor.

In some groups, such as the ferns, one can safely postulate effective wind distribution for a distinctly high percentage of both low- and high-altitude species found in the Pacific islands, whether one considers Samoa, Fiji, or the much more distant Hawaii, Tahiti, or the Marquesas Islands. Many fern species are common to the floras of these islands and to the vastly larger islands of Malaysia, while a considerable number of them are found in the tropics of both hemispheres. Here it seems probable that their microscopic spores may be carried by the wind for vast distances. Although in this group there are a large number of apparently wind-distributed species common to Malaysia and Polynesia, it should be noted in passing that the direction of the prevailing winds is not particularly favorable to an eastern distribution of wind-borne species from Malaysia; however, we know practically nothing at present regarding wind directions in the stratosphere, which may be very different from what they are at low medium altitudes.

LIMITATIONS ON WIND-DISTRIBUTED SEEDS

As contrasted to the ferns, the orchids show a very different picture or pattern of distribution. Because all of the species of this very large family produce myriads of minute dust-like seeds, it is assumed that this is an adaptation for dissemination through the medium of the wind. And yet this family throughout Malaysia, with at least 5,000 known species in that part of the world, is noted for the very high percentage of local endemism of its very numerous species. Thus, in the Philippines, approximately 900 orchid species are known, but only about 130 of them have been discovered outside of the limits of that island group. The situation in New Guinea is even more striking, with about 2,000 species of orchids now known; yet most of them are strictly limited to that one island.

COMPARISON OF THE MICRONESIAN-POLYNESIAN ORCHID FLORA WITH THAT OF MALAYSIA

Assuming that the orchids are wind-distributed, it is interesting to compare the known orchid flora of the various parts of the Pacific basin. In the Japanese-mandated islands, representatives of 42 genera and about 80 species are known; from Fiji about 45 genera and 125 species; from Samoa about 44 genera, 115 species; from Tahiti 16 genera and 30 species; from the Marquesas 3 genera and 4 species; and from Hawaii 3 genera and 3 species. Here we have a striking illustration of the inability of orchid species to extend their range over wide expanses of ocean, and an even more striking example of the attenuation of the Malaysian flora as one progresses to the east in the Pacific basin; for the Polynesian orchid flora is a Malaysian one, as to its constituent genera. It thus seems to be rather unlikely that wind has been the determining factor as far as the orchids are concerned, and one may assume this to be the case for certain other natural families of plants as well. It may be that the dust-like seeds of the orchids are distributed by the wind over great distances; but if so, they either lose their germinating power, or if on reaching favorable habitats they do germinate, the plantlets fail because their roots lack the opportunity of coming in contact with the fungi known to be essential to the growth and development of these plants, which are mostly epiphytic. Another factor that must be considered in this group is the remarkable adaptations to pollination by specific insects; the essential insect may be absent in the regions where the orchid flora is so very limited, for without the presence of the proper insect, actual pollination never or but very rarely takes place.

ATTENUATION OF MALAYSIAN GENERA IN THE PACIFIC BASIN

In genus after genus common to Malaysia and the islands of the Pacific, be they very large or medium-sized ones, this marked attenuation holds in the number of species found in any of the groups of high Pacific islands, as compared with the very large numbers found on the larger islands to the west. Very recently the same thing has been shown to be true for various groups of insects.

EXAMPLES OF ATTENUATED DISTRIBUTION

Elæocarpus may be accepted as typical (fig. 255). This is strictly an Old World genus, with between 400 and 450 recognized species. There are a few species in India, Ceylon, Madagascar, and tropical Africa, but within the area considered in figure 255, extending from southern Japan southward to Tasmania, and westward to Siam, Burma, and Sumatra, there are at least 400 different species already known. Malaysia as a whole can only be considered as the center of origin and development for this particular genus. The attenuation of this large genus

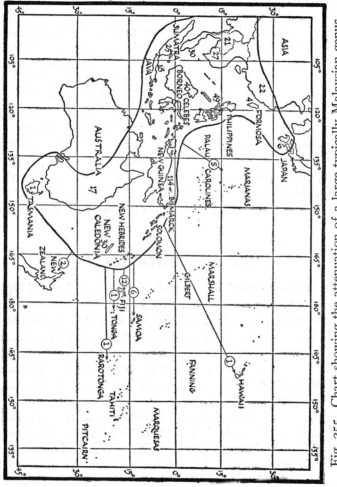

Fig. 255. Chart showing the attenuation of a large typically Malaysian genus, *Elaeocarpus*, in Polynesia.

as one proceeds eastward in the Pacific basin is very striking, with only five species in the Palau-Caroline-Marianas groups, twelve in Fiji, six in Samoa, two in New Zealand, and one each in Tonga, Rarotonga, and Hawaii. For some reason the genus failed to reach the distant high islands of Tahiti, Marquesas, and Pitcairn, although in these islands the soil and climatic conditions are favorable to the growth of species of this genus; they are not adapted to the conditions found on the typical low islands of the Pacific basin. This very striking attenuation of the larger Malaysian genera in the Pacific basin is characteristic, whether we are dealing with *Elæocarpus, Ficus, Syzygium (Eugenia), Canarium, Calamus, Calophyllum, Medinilla, Macaranga, Antidesma, Glochidion, Sterculia, Dendrobium, Bulbophyllum,* or any other natural group. In the map (fig. 255), the figures within the solid line indicate the approximate number of species of *Elæocarpus* at present known from each geographic area.

VARIOUS HYPOTHESES

What is the explanation? Were the floras of the isolated high Pacific islands once much richer than is now the case, and have numerous species ceased to exist? This is highly improbable, for it is very unlikely that these insular floras were ever richer than they are today; in fact, they now support many more genera and species than they did a century or two ago, for many man-distributed plants have been introduced and naturalized within that period.

Was there once a continuous land mass in what is now the Pacific basin? Most geologists dismiss this as impossible. From a strictly botanical standpoint it could be reasoned that were these widely scattered high islands the remains of a former land mass, then certainly the floras of the existing islands would be much richer than is actually the case—that is, assuming that they have been continually above sea-level from early geologic times, which is perhaps assuming too much. Have there been in the past more extensive archipelagic conditions? Has there been a general depression of most of these hypothetical island groups?

The plausible but not very logical idea of former land bridges between parts of what is now Malaysia and parts of what is now Polynesia, has been proposed and vigorously supported; hypothetical land bridges have been scattered right and left all over the Pacific basin to explain the present-day distribution of this or that group of plants. It may be that the explanation is to be sought in a series of island arcs—more or less regular rows of islands perhaps greater in extent in former geologic periods, separated by deep troughs—and that the possible former land areas have become more and more reduced through changes in the earth's crust; or again that the process of mountain growth is still active in certain parts of the Pacific basin. Were the latter the case, it would then be clear that we could expect only poor floras on relatively recent mountainous islands in the Pacific. There is much geological evidence now available indicating that this may be the true explanation of the situation in the western Pacific region, appertaining especially to those groups of islands extending southward from Japan to the Marianas, Caroline, and Palau Islands.

Whatever the factors may have been that govern the distribution of plants in the widely scattered island groups of the Pacific basin, we cannot say with certainty or finality, but the fact remains that representatives of various plant genera did, in the past, reach these now remote islands; and their descendants have persisted, largely as endemic elements, regardless of what has happened in the interim. The present-day biological alliances of the remote Pacific islands are, however, overwhelmingly with the floras and faunas of the great islands of the western Pacific.

13
The Significance of Certain Local Plant Names

The number of more or less specific local plant names in the Malaysian, Melanesian, Micronesian, and Polynesian regions is very great. A peculiarly high percentage of them, within the geographic areas wherein they are in current use, is definitely fixed in that they are, in general, applied to individual, more or less characteristic plant species. They include, of course, those names used to designate strictly native species, as well as those applied to introduced, cultivated, and naturalized ones. At the present time, within the vast area designated, there are certainly in excess of 45,000 different native plant names already recorded in association with their equivalent Latin binomials. At the same time, there are many thousands of currently used names, more perhaps of local as distinguished from wide geographic use, that have never been recorded.

PLANT NAMES AND LOCAL LANGUAGES

The total number of these names is not surprising when one considers that in excess of 500 different languages or dialects are spoken by the native peoples of the region covered, extending from northern Sumatra to the Philippines, southward and eastward to New Guinea and its adjacent islands, and across the Pacific to Hawaii and the Marquesas Islands. Another reason for the large number of names is the fact that widely distributed species, particularly those of some economic importance, may have in excess of 100 or more different names within their natural geographic range. The enormous development in genera and in species characteristic of the larger islands and island groups is a further reason, together with the fact that the percentage of local endemism is very high, as indicated in the chapter on the primary forest. Many of these local names are in current use only in very restricted areas and among small groups of people speaking a common language or dialect. Others extend with minor variants or cognate forms over vast distances —in some cases all the way from Madagascar to Hawaii, applied to the same plant species.

The remarkable thing about this diversity is the relative fixity of individual plant names, at least within restricted areas, in that a given name is always or usually applied to a single species. Sometimes, of course, a name is used more or less in the sense of a generic designation and is employed indiscriminately for a large number of different species within a natural group, family, or genus that have certain manifest qualities in common. Thus in the Philippines the word *dápo* is used to designate most species of epiphytic orchids, as well as other unrelated epiphytic plants; and the word *baléte* is applied to a great number of different species of *Ficus*, but always to those in that group which start as epiphytes—the banyan type or strangling figs that eventually strangle the host plant on which they developed as young plants.

Most of these names naturally form a part of the original language of specific peoples, various branches of the Malayan, Melanesian, Polynesian, Micronesian, and other ethnic groups; many of them are descriptive. Yet at the same time there is often a very wide and relatively fixed use of a single name for individual species, including some that are of strictly natural distribution and others where the individual plant species have been distributed by man himself. In general, when one compares a long series of Polynesian plant names with a series of Malaysian ones, he is impressed with similarities that occur over and over again, but with one rather marked difference. In the Polynesian languages the words are softened by the elimination of certain consonants. Thus the Philippine *níog* (coconut) becomes the Polynesian *ni, niu, niue,* and the Malaysian *bágo,* applied to various species having string bast fibers, appears in Polynesia as *bau, hau, fau,* and *vau.* This elimination of certain consonants is a genetic difference between the two language groups.

WIDE USE OF THE SAME NAME

Sometimes a widely distributed plant name may be restricted to a single species or to various representatives of the same genus, but at other times it may be applied to totally unrelated ones. In the latter case the different plants always have certain uses in common. Thus, for example, is the word *túba* applied to various representatives of the leguminous genus *Derris,* all woody vines, representatives of this genus being the most efficacious of all Malaysian plants used for stupefying or poisoning fish. This name is in use all the way from Sumatra to the Philippines and even appears as far to the east as Fiji in the cognate form *dúva.* Yet *túba* is also widely applied to an euphorbiaceous erect shrub, the croton-oil-plant (*Croton tiglium*), not because there is any similarity between *Derris* and *Croton,* but merely because macerated parts of *Croton tiglium* are also efficient in suffocating or poisoning fish.

POSSIBLE ORIGINS OF CERTAIN NAMES

In this same category is the word *bágo.* This name, or manifestly cognate forms of it, is applied to several totally different plants in unrelated natural families, not because of any similarity in the plants themselves but because all of them, whether shrubs, trees, or vines, yield particularly strong bast fibers and are thus of economic importance. This word is used over an even wider geographic range than is *túba.* It may be assumed that the basic form was *bágo,* and this in Malaysia and the Philippines is the cognomen of the widely distributed and sometimes planted *Gnetum gnemon,* a small tree, and other climbing species of the same genus. The bast fiber is particularly strong, and whenever possible it is sought when a bowstring is to be made. But *bágo, bálibágo, málabágo, págo, palúpo, báro,* and the softer forms in Polynesia, such as *hau, fau, bau, vau,* are used to designate the entirely different malvaceous *Hibiscus tiliaceus,* a tree with broad heart-shaped leaves and large yellow flowers; it too yields an important bast fiber. Again the form *bágo,* with variants, is applied to another series of shrubs and small trees in the Indo-Malaysian genera *Wikstrœmia, Phaleria,* and in Indian even to *Daphne,* all with strong bast fibers; all of these belong in the Thymelæaceæ. It is suspected that the word *bágo* originally meant bast, even as the American basswood (*Tilia*) was apparently modified from bastwood. At any rate, over vast distances and among widely separated peoples speaking greatly diversified languages, all the way from the Mascarene Islands and India to the far reaches of the tropical Pacific, the same word, often more or less modified, is now applied to a series of totally unrelated plants, all of which, however, have one common denominator in that they all yield strong bast fibers.

WIDELY USED NAMES FOR CULTIVATED SPECIES

In a different category is a series of very similar names applied to cultivated plants, such as the coconut palm, a species that has manifestly been to a very large degree disseminated by man as opposed to distribution through natural agencies. This palm is clearly one of Old World origin, although now it is also widely distributed in tropical America. There are actually myriads of different names for the coconut and its recognized varieties in the Old World tropics; for like all cultivated plants, it is variable, and the variations are reflected in the size of the plant, as well as in the size, color, and other characters of its fruit. In the Malaysian region we find the *halámbir-kalámbir* series of names in the Malay Peninsula, Sumatra, and Java; the *kalápa* series in Java and Borneo; and in the Philippines the universally used name *niog*; outside of the archipelago we find the *pol* series of Persian origin, the *ong* series of Ceylonese origin, and others.

Now consider for a moment the *nia-niu-niue-niog* series, for these names are manifestly cognate forms of a single original name. These names, in one form or another, are in actual use all the way from Madagascar to Hawaii, and are always strictly applied to the coconut palm. Because of the very wide geographic range, as compared with the geographically very limited ranges of the other series of names mentioned above, we may safely assume that the *nia-niog* series represents a much older name than the others. And since the coconut palm is essentially a man-distributed species, we may also assume that the name was transmitted with the palm itself, by man, as he extended over the Old World tropics, carrying this very important species with him in his travels. And when it was finally introduced into tropical America by the Portuguese and the Spaniards, it came in under its Portuguese name, *cóco,* which in turn was apparently of Malayan origin.

Another excellent illustration of the distribution of the name of the plant with the plant itself is illustrated by the greater yam (*Dioscorea alata*), now universally cultivated in the tropics of both hemispheres but of Old World origin, and most certainly for the most part man-distributed. Throughout Malaysia from the Malay Pensinula to the Philippines and New Guinea, we find such names as *uébi, úbi, úwi,* and *húwi* widely used for this cultivated species, and throughout Polynesia as far to the east as Hawaii and the Marquesas Islands this same name appears in such forms as *úi, úfi, úhi,* and *pahúi.*

HOW PLANT NAMES WERE SPREAD

This phenomenon of the dissemination of plant names with the plants that are largely man-distributed, like the coconut palm and the yam, is a common one, particularly for agricultural, ornamental, medicinal, and other economic species. Here it is pertinent to interpolate the evident but usually not appreciated fact that before the discovery of America by Columbus in 1492, there was not a single cultivated basic food plant common to the two hemispheres. And yet today we find very many plants of American origin cultivated, and many of them naturalized, in the Old World tropics, even as we find many species of Old World origin now established in the American tropics. The reason for this is very simple. Just as early man, after he had established even the simplest type of agriculture, took with him the food plants with which he was familiar at home when he occupied new areas, so in more modern times civilized man has done exactly the same thing, although on a much larger and very much wider scale. The situation now, as compared with the period antedating the expansion of the European colonizing nations, has become distinctly more complex, for modern man is the most important single factor in the actual dissemination of both plants and animals.

AMERICAN *VERSUS* EURASIAN CULTIVATED PLANTS

As the Eurasian cultures were built up over the course of the centuries on the basis of an agriculture founded wholly on plants and animals native to some part of the eastern hemisphere, so in turn the pre-Columbian cultures in America were also developed and supported by a series of strictly native American plants and animals that were domesticated by early man in America. Without agriculture there could be no great developments in culture and civilization, for a constant dependable food supply is basic. There is no evidence that early man entered America with an established agriculture, but there is a vast accumulation of data indicating that once in America, and once in contact with various native species that were eventually found to be susceptible to domestication and cultivation, wholly independent of any Old World influences, he established a new type of agriculture—new insofar as it was based on a new series of plants. On the basis of this agriculture, he gradually developed the high types of civilizations in Mexico, Central America, and parts of South America that amazed the early European explorers. In ultimate analysis we do not have to look to the Old World for the influences that established and developed the pre-Columbian civilizations in America, for they were essentially a growth from the American soil.

Just how strong these differences are, in terms of cultivated plants and domesticated animals, is evidenced by the statement that before the time of Columbus not a single basic cultivated food plant and not a single domesticated animal, other than the dog, was common to the two hemispheres. All of the cultivated cereals, with the single exception of maize or Indian corn, are natives of and were domesticated in Eurasia, including wheat, rye, barley, rice, sorghum, oats, millet, and others of minor importance. At the same time, most of the commonly cultivated vegetables in temperate regions are also natives of Europe or Asia, as well as all of the temperate-zone cultivated fruits, such as the apple, pear, peach, plum, cherry, and others. With few exceptions, and these of minor importance, our domesticated animals also came from the Old World.

Unknown in Europe or in Asia until after the voyages of Columbus and Magellan were the following important food plants, all contributions from the Americas: maize or Indian corn, potato, sweet potato, cassava or tapioca, all of the garden beans, Lima bean, squash, pumpkin, tomato, pepper, sunflower, Jerusalem artichoke, and various others of less importance, while in the animal kingdom about the only American contributions were the guinea pig, llama, alpaca, turkey, and Muscovy duck.

The early European explorers and colonists transmitted the Eurasian agricultural plants and domesticated animals to America, and at the same time they introduced the important American economic series into the Old World, thus initiating the modern development of man as the most important single agent in reference to the distribution of plants; this is a development within the past 450 years. Thus the expanding European peoples in colonizing recently discovered lands accomplished exactly what prehistoric man did on a smaller and much more limited scale. Both carried with them the plants and animals on which they, in part, depended for food. The essential difference is that early man was limited to the one hemisphere or the other, being in general unable to cross either the Atlantic or the Pacific Ocean, but modern man solved the problem of finding his way across both oceans and back again.

PLANT NAMES IN THE PACIFIC REGION

Returning to the problem of plant names in the Pacific region, it is clear that most of these names are of local origin, and further, that very many of them are of very local use. They were apparently originated by this or that ethnic group as the plants became familiar to them, or as individuals discovered potential economic uses of this or that species. Migratory or expanding peoples familiar with a plant species in the Sunda Islands would naturally apply the same name to the same species if they observed it in new areas that came under their control, such as the Philippines, Lesser Sunda Islands, Celebes, etc. If, however, they discovered a plant species with which they were not previously familiar, the only resource they had was to coin a new name for it, and this they apparently did on a large scale.

EXOTIC PLANT NAMES IN THE PACIFIC REGION

Within Malaysia and the Philippines there are several striking series of plant names, many of them now in very wide use, which have nothing to do with aboriginal Malaysian plant names. They are applied to various introduced and naturalized species, all of which are known to have been introduced into the archipelago within relatively recent times. It is worth while to glance at the early history of the archipelago. We know that the advanced Indian peoples, as well as the Chinese, have had more or less close contacts with the Malaysian region for 2,000 years or perhaps longer. As the Indian peoples commenced to colonize Sumatra and Java and to extend their control over the more distant parts of the archipelago, they naturally introduced the economic plants with which they were familiar in India. Thus we find today in extraordinarily wide use in the archipelago a long series of unmodified or slightly modified Sanskritic names which, with the plants themselves, originated in India. And yet today among the peoples of the archipelago at large there is no knowledge of this former domination of the region by the people of India. The general outlines of the histories of the great empires of Sri-Vishaya and Madjapahit are well known; these extensive empires dominated Malaysia previous to the Mohammedan influences that commenced to make themselves felt in the fourteenth century.

SANSKRITIC NAMES

As these invading Indian peoples brought with them their own culture, religion, and agriculture, naturally they supplemented the aboriginal agriculture by importations of additional species from India. Only a few of these numerous Sanskritic names can be cited here, but these will illustrate the point: *lasóna,* the onion; *malísa,* black pepper; *kachúmba,* the safflower; *kastúli,* the musk mallow (*Abelmoschus moschatus*); *kápas* or *gápas,* cotton; *pána* the jakfruit (*Artocarpus heterophylla*); *champáka,* the magnolia-like *Michelia champaca*; *lagúndi,* the chaste-tree (*Vitex negundo*); *gandasúli,* the ginger-lily (*Hedychium coronarium*); *dáua,* millet (*Setaria italica*); and *karambódja* for the watermelon.

CHINESE NAMES

The Chinese influence is evident, but since the Chinese came essentially as traders rather than as colonists, the number of plant names of Chinese origin is naturally smaller than of Indian origin. One readily recognizes Chinese forms in such plant names as *bátao* for the hyacinth bean (*Dolichos lablab*); *sítao* for the cowpea (*Vigna sinensis*); *kúchai* for the small onion (*Allium tuberosum*); *tungháo* for *Chrysanthemum coronarium*; *uńgsoi* for *Apium graveolens*; *péchai* for Chinese cabbage (*Brassica pekinensis*); and *kangkóng* for *Ipomœa aquatica.*

MEXICAN NAMES

In more modern times we have an exact parallel with these early introductions from Asia through the Spanish influence in colonizing the Philippines; and it should be remembered that up to the time of Mexican independence (1821), the Philippines were governed as a dependency of New Spain or Mexico. From shortly after the establishment of Spanish rule in the Philippines in the latter part of the sixteenth century up to about the time of the Mexican independence, the Spanish contacts with the Philippines were almost wholly via Mexico, and the Pacific leg of this long route was the long-continued Acapulco-Manila galleon system. It was only natural that the Spanish colonists and officials transmitted numerous economic species from Mexico to the Philippines, and vice versa. Since these native American economic species had no Spanish names, the Spaniards naturally adopted various local names as used in Mexico; and thus today we find a long series of plant names of Aztec origin unchanged or but slightly modified actually in use in the Philippines, introduced from Mexico with the plants themselves over a period of about 250 years. A few examples will suffice: *kamóte* for the sweet potato; *kalachúchi* for the frangipani (*Plumeria acutifolia*); *kakuáte* for the *madre de cacáo* (*Gliricidia sepium*); *kamanchíli* for the Manila tamarind (*Pithecellobium dulce*); *kamáte* for the tomato; *chocoláte* for cacao; *aposótis* for *Chenopodium ambrosioides*; *chíko* for the sapodilla (*Achras zapota*); *achuéte* for the annatto (*Bixa orellana*); and *kolítis* for various species of *Amaranthus*.

NAMES AS INDICATORS OF PREVIOUS CONTACTS

One does not have to be a comparative philologist to realize, in the series of Sanskritic and Mexican names, as with those of Chinese origin, that genetically these have little or nothing to do with plant names of Malaysian origin. The cited series form excellent illustrations of prehistoric and historic influences that must be taken into consideration in any intensive study of languages and customs. The several series of plant names reflect certain characteristics of the languages from whence they originally came—the Sanskritic group of India, the Chinese of southern and eastern Asia, and the Aztec group of Mexico, languages totally different from the very large number spoken in the Pacific area; and yet all have made their permanent contributions in the form of plant names to the living languages of the Pacific islands of today.

PLACE NAMES APPLIED TO PLANTS

Place names within and without the archipelago play their part. Thus the Manila tamarind, a plant of Mexican origin, is the name used for *Pithecellobium dulce* in India, for the tree was introduced into India from Manila shortly after the middle of the eighteenth century. In Java the tropical American sapodilla (*Achras zapota*) is known as *sáwo manila,* the peanut as *kátjang manila,* and the soursop (*Annona muricata*) as *langka manila.* And while in the Philippines the American *Cassia alata* is known as *kapúrko* or *akapúlko,* for it was originally introduced from Acapulco, Mexico, it in turn is known in Java as *ki manila.* The word "manila" in these and other cases points to the place or origin for the early introductions into Java. Various other well-known geographic names as Malacca, Timor, Bali, Java, Bandung, Ambon, Cambodia, China, and others are frequently noted as forming parts of plant names in Malaysia, indicating a belief—not always true, however—that the indicated geographic place name had something to do with the place whence a certain plant originally came.

THE RELATIVE FIXITY OF NATIVE PLANT NAMES

Thus it is that when we come to deal with such a tremendously developed flora as that of the larger Pacific islands, we are forced to indicate species by the use of Latin binomials, for these, at least among botanists, are reasonably well fixed. But at the same time we are faced with thousands of native names of one type or another, and these in turn cannot be wholly ignored. The individual interested in particular plants, such as those yielding important fruits, gums, resins, or lumber, cannot avoid the use of local names; for in most localities these names are the only ones in current use, and doubtless will remain as such for centuries to come.

One might observe in passing that sometimes a well-established native name is really more definite and more fixed in application in such a region as Malaysia as a whole than are our vaunted Latin binomials; for well-established native names do not change through the centuries, while in some groups of plants, because of confusion in the early development of botanical classification, names long in use for a particular species will be dropped when it is found that an earlier published one actually applies to the same species. Native plant names are thus not to be ignored, no matter if the number in actual use is very, very great, and no matter if at first these impress one as meaningless. A botanist familiar with the general facts of plant distribution may even suggest that comparative philologists and anthropologists may find much of interest and value in a consideration of these plant names, for they have in the past been too generally overlooked and ignored by all groups, including the botanists. We should not be misled by the very loose application of our common English plant names at home, for among more primitive peoples the plant names in current use are too important to be ignored, since such peoples as a rule are in much closer contact with plants than are more advanced peoples.

14
Notes on Specific Islands and Island Groups

The preceding chapters in this volume have dealt in a rather general manner with some of the aspects of plant life in the islands of the tropical Pacific, as well as those of Malaysia as a whole. In general, the sketches of such plant formations as the primary forest, the secondary forest and open grass-land, the mangrove swamp, and the strand vegetation, will apply equally well to even wider areas, as in tropical southern Asia. However, in addition to the tropical regions, aside from the American borders of the Pacific basin, there are temperate regions such as the Aleutian and Kurile Islands and Japan that warrant brief consideration, as well as certain island groups nearer the American shores, and scattered ones here and there in the Pacific Ocean. Some of these are below briefly treated, proceeding from the Aleutians in the north, southward along the western border of the Pacific, with at the same time some mention of island groups as far to the east as the Galápagos. For complete bibliographic data on what has been published regarding the vegetation of the vast area extending from Tibet to Kamchatka and the Kurile Islands southward to Yunnan, Hainan, and Formosa, see the Merrill-Walker *Bibliography of Eastern Asiatic Botany,* pp. xliii, 719, 1938, which contains about 21,000 author-entries.

THE ALEUTIAN ISLANDS

These bleak rugged islands support a relatively scanty vegetation. Trees are entirely absent, but some small shrubs occur representing such groups as the willows (*Salix*), the birches (*Betula*), and small plants of woody or semi-woody representatives of *Ledum, Rhododendron,* and allied genera, particularly on those islands nearest to the mainland. The highest peak is 9,387 feet high. The total flora consists of only 480 species, a high percentage of these being grasses and sedges. The islands stretch over a distance of about 1,200 miles. The flora is remarkably poor when one considers the richness of the vegetation of warmer and more favorably located regions of similar extent. There are some distinctly showy flowers, for the plants develop rapidly in the short growing season. Such groups as lady-slipper (*Cypripedium*), iris (*Iris*), marsh-marigold (*Caltha*), globe-flower (*Trollius*), monkshood (*Aconitum*), windflower (*Anemone*), poppy (*Papaver*), fireweed (*Epilobium*), primrose (*Primula*), gentian (*Gentiana*), painted-cup (*Castilleja*), bluebells (*Campanula*), groundsel (*Senecio*), and others are represented. The species are, for the most part, those of very wide northern distribution. Our present knowledge of this limited flora is admirably summarized in Hultén's *Flora of the Aleutian Islands and Westernmost Alaska Peninsula with Notes on the Flora of the Commander Islands*, pp. 1–397, illus., 1937.

In general, the Aleutian flora is not very different from those of adjacent parts of Alaska to the east and Kamchatka Peninsula to the west, although, of course, the forest developments in parts of both of the latter regions are strikingly different from the treeless islands of this long chain. The Kamchatka flora may be readily compared by consulting Hultén's recent *Flora of Kamtchatka and the Adjacent Islands*, 2 vols., illus., 1926–1930, and Komarov's *Floræ Peninsulæ Kamtchatka*, 3 vols., illus., 1927–30. The total Kamchatka flora is limited to about 800 species, in spite of the much larger land area involved as compared with the small scattered islands of the Aleutian chain. In the Aleutians there has naturally been some intermingling of the Alaskan and the Kamchatkan floras.

THE KURILE ISLANDS

Extending to the southwest from the Kamchatka Peninsula for a distance of about 700 miles stretch the Kurile Islands, in the direction of Japan. These rugged islands with altitudes up to 7,783 feet are bleak to a large degree, and like the regions to the north and northeast, they also support a poor flora. There are, however, some forested areas. The total land area is about 4,000 square miles. Modern lists of plants are available for some of the islands such as Paramushir in the extreme north (Kudo, 1922), 284 species; Shikotan (Ohwi, 1932–33), 527 species; Shimashu (Tatewaki, 1933); and in the somewhat older and basic work of Miyabe (1890) on the flora of the entire group. To the west and adjacent to the Asiatic coast lies the long island of Sakhalin, 24,560 square miles, with peaks up to about 5,000 feet high. There are extensive forested areas and much tundra. Its flora is well known through Sugawara's *Illustrated Flora of Saghalien*, 4 vols., with 892 plates, 1937–1940; 1,166 species are known from that island.

JAPAN

The Japanese flora is an exceedingly rich one, for it contains not only numerous endemic elements, but also many northern and Asiatic types and in the south a distinctly large infiltration of Malaysian elements—species that have apparently extended northward along the Asiatic continental shelf, because of favorable climatic conditions, to the southern part of this island group. In excess of 4,000 species of ferns and flowering plants are now known from Japan proper, and if the Riu Kiu Islands and Formosa be included, then many hundreds of additional ones. The flora is so rich that this is scarcely the place to consider it in any detail. In addition to the numerous earlier works on the vegetation of Japan, in modern time Japanese botanists have issued various manuals and descriptive floras, and in general the flora has been unusually well illustrated in such works as Makino's *Futsu shokubutsu dzufu* (Illustrations of Common Plants), 5 vols., 1912–13, Murakoshi's *Nai-gwai shokubutsu genshoku dai-dzukan* (Illustrated Encyclopedia of Botany), 13 vols., 1935, his *Dai shokubutsu dzukan* (Comprehensive Illustrations of Plants), 1932, and *Genshoku dzusetsu shokubutsu daijiten* (Comprehensive Botanical Dictionary in Color), 1938,

NOTES ON SPECIFIC ISLANDS

and Terasaki's *Nippon shokobutsu dzufu* (Illustrations of Japanese Plants), 1933, and its Supplement, 1938. Terasaki's work contains good illustrations of 4,000 species of plants.

One phase of the Japanese flora, as well as that of eastern Asia in general, is worthy of special mention. This is that from the standpoint of plant geography these regions show a series of remarkable alliances with eastern North America, rather than with the western parts of the American continent. In genus after genus, living representatives are known only from eastern North America and Japan and eastern Asia. Examples are our common mayflower (*Epigæa*), hickory (*Carya*), moonseed (*Menispermum*), silver-bell (*Halesia*), tulip-tree (*Liriodendron*), sweetgum (*Liquidambar*), coffee-tree (*Gymnocladus*), yellow-wood (*Cladrastis*), witch-hazel (*Hamamelis*), sassafras (*Sassafras*), and many others. Sometimes a single species is involved; at other times two or more closely related ones. From a strictly geographical standpoint, one would surmise that the flora of Japan and of eastern Asia would be most closely allied to that of western North America; there are, of course, such relationships.

To explain the problems of discontinuous distribution involved in these "islands" of vegetation separated by half the circumference of the world, geologic history holds the clue. The great extensions of the ice caps southward in eastern North America, as far as Long Island, in the geologic periods preceding the present, eliminated the vegetation over vast areas in North America. As the ice sheets extended southward, various species representing what was then a fairly continuous flora extending from eastern Asia across northern North America to its eastern seaboard, also migrated southward; as the ice sheet retreated northward in America, at the close of the glacial period, the vegetation also followed it, but failed, because of persistent adverse conditions in the north, to re-establish itself over vast parts of northern North America. Thus we find today these numerous relic species of what was once a continuous flora occupying widely separated lands—eastern North America and eastern Asia.

THE RIU KIU ISLANDS AND FORMOSA

In general, as we proceed southward from Japan proper, more and more subtropical or even tropical elements appear, derived from the warmer parts of Asia and from Malaysia. Their limit of northward extension was certain climatic factors, chiefly temperature, just as the northern types that extend southward are, for the most part, limited by climatic conditions, being found only on the higher mountains in the tropics where low temperatures prevail. The Riu Kiu flora, as would naturally be expected, is composed of Japanese and Formosan elements intermingled with certain Chinese types. Formosa itself, with the extreme altitude of 14,720 feet in its Mount Morrison (Niitikayama), presents a most intriguing assemblage of Asiatic and northern types, as well as many tropical ones, as would be expected from its geographic position, its altitudinal range, its geologic history, and its location on the edge of the Asiatic continental shelf. And yet its flora is essentially continental (Asiatic) as compared with the insular (Malaysian) floras of the great islands to the south and southwest. The dividing line extending northward along the west coast of Luzon and thence eastward into the Pacific close to the southern tip of Formosa and between that large island and the contiguous small one of Botel Tobago, is fairly definite. Very few strictly Philippine types reach Formosa, but a considerable number occur on the little island of Botel Tobago; and yet from Y'ami, the most northern island of the Philippine group, on a clear day one can see the Formosan coast. Our present knowledge of the Formosan flora, which approximates 4,000 species, is well summarized in Hayata's *Icones Plantarum Formosanarum*, 10 vols., 1911-21.

HAINAN

This island, located across a shallow narrow strait separating it from the tip of the Luichow Peninsula, also lies on the continental shelf. Its flora is essentially tropical, consisting of approximately 2,500 known species. Its mountains are not nearly so high as those of Formosa, and yet they are sufficiently rugged, ascending to an altitude of about 6,200 feet. Its flora is made up of continental and chiefly Chinese elements, supplemented by many otherwise known only from Indo-China, together with many Malaysian representatives that failed to reach China proper or Formosa; many of these also occur in the Philippines. Our present knowledge of the flora is summarized in Merrill's *Enumeration of Hainan Plants,* 1927, his as yet unpublished *Working List of Hainan Plants,* and Tanaka and Odashima's *Census of Hainan Plants,* 1938.

THE PHILIPPINES AND MALAYSIA IN GENERAL

Within the present century the Philippine flora has been intensively studied. It is exceedingly rich in both genera and species, with a very high percentage of specific endemism. On the higher mountains one finds representatives of many northern groups, such as pine (*Pinus*), yew (*Taxus*), *Aster,* goldenrod (*Solidago*), violet (*Viola*), *Anemone,* buttercup (*Ranunculus*), rue (*Thalictrum*), gentians (*Gentiana*), raspberries (*Rubus*), rose (*Rosa*), blueberries (*Vaccinium*), *Rhododendron,* lily (*Lilium*), honeysuckle (*Lonicera*), and numerous others. The lowland flora is strictly tropical and is made up very largely of characteristic Malaysian genera, those usually well represented either all over the Malay Archipelago and tropical southern Asia, or others characteristic of the Sunda Island group (Borneo, Sumatra, Java), or of the islands to the south and southeast, such as Celebes, the Moluccas, New Guinea, and the Solomon Islands. It is of interest to note that as material from New Guinea and its adjacent islands becomes available for study, more and more representatives of groups originally described from Philippine material appear there. At both low and high altitudes in the Philippines, there is a peculiar infiltration of strictly Australian types, much stronger here than in western Malaysia, but none of them extend to Formosa, although some reach as far north as the small islands between Luzon and Formosa. Again, on the higher mountains, where also certain strictly Australian types occur, one finds what may be interpreted as the most eastern and, in some cases, the most southern extensions of certain Himalayan types. These strictly Asiatic (continental) types in the Philippines include not only high-altitude species but certain low-altitude ones as well.

Generally speaking, the generic constituents of the various types of vegetation in the Philippines are the same as those in Malaysia, with, however, representatives of certain typical Asiatic and Australian genera that failed to reach western Malaysia. The constituents of the second-growth forests are likewise the same, for the most part, even as to the species, as those that occur in similar areas in the Malay Archipelago as a whole; this same statement applies to the mangrove swamp and strand vegetation, to the characteristic natural open grass-lands, to the weeds of cultivation, and to the plants cultivated for ornamental or economic purposes. Within the Philippine primary forests, however, the picture is very different, for most of the species that occur here are known only from that archipelago, although others are widely distributed; most of the genera there represented are Malaysian. As noted elsewhere, the percentage of endemism in the primary forests of the Philippines approximates 84 per cent. In general, about this same percentage of endemism applies to all of the large islands of the entire Malaysian region such as Sumatra, Borneo, Celebes, and the lesser islands in the region south of the Philippines, and to New Guinea, the Solomon Islands, New Hebrides, and New Caledonia, as far as the vegetation of the forested areas is concerned. This entire insular region, by far the largest archipelago in the world, is thus a most fascinating place to study problems of geographic distribution and is of equal interest from the standpoint of purely descriptive systematic botany. The Malay Archipelago as a whole, including the Philippines and New Guinea, and its contiguous islands, supports one of the very richest floras of the world.

Our present knowledge of the very rich Philippine flora, with over 8,500 known species of flowering plants, to which approximately 1,000 species of ferns and fern allies should be added, is reasonably well summarized in Merrill's *Enumeration of Philippine Flowering Plants,* 4 vols., 1923–1926.

GALAPAGOS ISLANDS

This group, located relatively close to the coast of Ecuador, has long been noted for its striking endemism both of plants and of animals. The botanical alliances are almost wholly with neighboring parts of South America. The lower slopes of all of the islands are sterile and dry, but following rains desert annuals spring up, quickly produce flowers, mature their seeds, and disappear. The perennial growth in such areas is largely composed of cacti and small-leaved shrubs and other plants. The more elevated islands are moist at higher altitudes, and where moisture prevails the flora is much richer than in the desert areas. Many of the shrubs and trees are spiny; such genera as *Mimosa, Acacia, Parkinsonia, Castela, Zanthoxylum,* and others are represented. About 60 different species of ferns are recorded. Epiphytes, largely ferns and bromeliads, occur only at higher altitudes and are neither abundant nor showy. Approximately 600 different species of plants make up this flora, of which about 40 per cent are known only from this island group. For detailed information see Robinson, B. L., *Flora of the Galapagos Islands,* Proc. Amer. Arts & Sci. 38:77–269, illus., 1902 (Reprinted as Contributions from the Gray Herbarium No. 24), and Stewart, A., *A Botanical Survey of the Galapagos Islands,* Proc. Calif. Acad. Sci. IV. 1:7–252, illus., 1911.

HAWAII

Like other remote high islands in the Pacific, the Hawaiian Islands have a most interesting indigenous flora, characterized by a very high percentage of endemism. In general, the dominant genera are Malaysian, with some Australian types, but some are manifestly allied to American groups. The introduced element is very large, including economic and ornamental species purposely introduced and many hundreds of different kinds of weeds and weedy plants, which have become naturalized. Many of the indigenous genera are represented by endemic species—or subspecies, varieties, forms, and perhaps hybrids—whose characters are not well fixed, whole groups apparently being more or less in a condition of flux. The result is a very large number of local types distinguishable only by minor characters; these are of strictly limited distribution. Plant distribution within the archipelago, as is the case with the land snails, is governed by local climatic conditions. Parts of the islands are very dry, while others are very wet; there are all degrees of gradation as between the constantly dry and the constantly wet regions. The great altitude of the higher mountains (13,835 feet on Hawaii), of course, varies the habitat and permits a considerable number of high-altitude species to thrive.

Much has been published regarding the constituent parts of the flora. The basic reference work is Hillebrand, W., *Flora of the Hawaiian Islands; a Description of Their Phanerogams and Vascular Cryptogams,* pp. i–xcvi, 1–673, 1888, supplemented by the recent Degener, O., *Flora Hawaiiensis, or the New Illustrated Flora of the Hawaiian Islands,* Books I–IV, 1932–40; the latter contains 400 illustrations with ample descriptions, including both native and introduced species. Fortunately there is an important local source of information in the Bernice P. Bishop Museum in Honolulu, with its excellent library and herbarium facilities.

OTHER SCATTERED PACIFIC ISLANDS

In general, the floras of the Pacific islands are reasonably well known, and the plants of a number of them have been treated in special works, particularly in the publications of the Bishop Museum, Honolulu. Thus available from the Bishop Museum are the basic lists or standard floras of various island groups and of individual islands, such as the Southwestern Pacific (Marquesas, Tuamotus, Austral, and Rapa), Brown in 1931 and 1935; Johnston and Wake, Christophersen in 1931; Samoa, Christophersen in 1935 and 1938, and also, published elsewhere, Rechinger in 1907–17, and Setchell in 1924; Flint,

NOTES ON SPECIFIC ISLANDS

St. John and Fosberg in 1937; Niue, Yuncker in 1943; Fiji, Gillespie in 1930 and 1932, and Smith in 1936, and, published elsewhere, Smith in 1942 and Gibbs in 1909; Makatea, Wilder in 1934; Rarotonga, Wilder in 1931, and, published elsewhere, Cheeseman in 1903; and Palmyra, published elsewhere, Rock in 1916. Various basic publications appertaining to individual islands of the region, published elsewhere than at the Bishop Museum, are Drake del Castillo on Tahiti, 1892; Hemsley on Tonga, 1894; Guillaumin on New Caledonia, 1909–1939; Guillaumin on the New Hebrides, 1931–34; Maiden on Funafuti, 1914; Safford in 1905, and Merrill in 1914 on Guam; and on all of Micronesia, Volkens in 1914, and the more recent publication, Japanese text, Kanehira, R., *Flora Micronesica*, pp. 1–3, 1–8, 1–468, 1–27, pl. 1–21, fig. 1–211, 1933. In general, full references and biblographical data regarding these publications and numerous others may be secured by consulting Merrill's *Bibliography of Polynesian Botany 1773–1935*, Bishop Museum Bull. 144: 1–194, 1937, and its unpublished Supplement. A selected bibliography of basic reference publications on the plants of the region covered by this book appears on page 248.

15

Notes on Botanical History, Exploration, and Bibliography

When European explorers first came to the Orient, their interests were largely commercial, and the lodestar, as far as Malaysia was concerned, was the very lucrative spice trade. The Portuguese reached India in 1498 and established their first settlement there two years later. Ten years afterward they arrived in Amboina, and in 1521 there established a permanent post; the Moluccas were an important center for the production of nutmegs, cloves, and other spices. The Spaniards reached the Philippines in 1521, but did not attempt colonization until nearly half a century later. At the end of the seventeenth century, the Portuguese colonies in Malaysia were taken over by the Netherlanders, and the Netherlands East India Company was established in 1602. Still later, the British entered the scene, and within the memory of living men (1884), the Germans made northeastern New Guinea (Kaiser Wilhelm Land) a German protectorate. Japanese entry into the scene dates from 1919 when Japan received, under mandate, the Micronesian islands that had been acquired by Germany following the Spanish-American War. The Portuguese made only very slight contributions to our knowledge of the natural history of the region. The first real efforts came from the Netherlanders and the British.

THE WORK OF RUMPHIUS

In general, in these early years of European colonization, commerce was the chief objective, and while certain individuals here and there were interested in natural history, it was not until well toward the close of the seventeenth century that any important contributions were made to our knowledge of the flora and fauna of the region. The basic work, and one that is still a great source of information regarding the plant life of Malaysia, was the monumental *Herbarium Amboinense* of Rumphius, for he, an employee of the Netherlands East India Company, assembled an enormous amount of information about both plants and animals. His classic work was completed about 1690 but was not published until 1741–55, when it appeared in six folio volumes illustrated by nearly 700 plates (see Merrill, E. D., *An Interpretation of Rumphius's Herbarium Amboinense*, 1–595, 1917). He has been well characterized as "the Pliny of the Indies," for he was one of the outstanding naturalists of all time.

RESUMPTION OF INTEREST

Following the death of Rumphius in 1702, there was a marked decline in interest in matters appertaining to natural history, although here and there some data were compiled and published. Little field work was accomplished, for the stimulus to field work, as we understand it, did not come until the publication of Linnæus' *Species Plantarum* in 1753. The simple system that he established for designating plant and animal species by binomial names, and his artificial but very simple system of plant classification, acted as a tremendous impetus, and soon the results became evident in increased field work and greatly increased publication.

Previous to 1753, various systems of classification of plants had been proposed, but these were unsatisfactory. The idea of the genus had become fairly well established by the end of the seventeenth century, but still the old system, dating back to classical times, of utilizing descriptive sentences to designate plant species remained in vogue. Thus the common and widely distributed coconut palm, consistently known as *Cocos nucifera* since this name was published by Linnæus in 1753, was well known to European botanists long before that date; it appeared in various works as *"Tenga," "Calappa," "Palma indica nucifera," "Palma indica coccifera angulosa,"* and *"Coccus frondibus pinnatis; foliolis ensiformibus margine villosis."* If single names or binomials did appear here and there in this early botanical literature, such simple names were purely accidental. If we complain about Latin binomials today, for both plants and animals, consider for a moment what scientists had to master before 1753.

EFFECT OF THE BINOMIAL SYSTEM

Soon after the binomial system was established in 1753, we note a quickened interest in the botany of the geographically better known parts of the region, beginning with Burman's *Flora Indica* (1768), in which he attempted to record what was known to him regarding the botany of both India and Malaysia. Shortly after this date, various technical periodicals were established; in them as well as in individual volumes published here and there in Europe and in England, appeared shorter or longer papers wherein plants from the Indies were described. Botany, however, did not come into its own for any considerable part of the region until well into the succeeding century, for here local interests of colonizing nations were involved. Then there commenced to appear the extensive series of publications of the plants of the Netherlands East Indies, Jack (1820–1825), Blume (1822–1859), Korthals (1830–1855, Junghuhn (1830–1863), Miquel (1838–1870), and others on the vegetation of Malaysia, as well as basic contributions to our knowledge of the floras of the Malay Peninsula, the Philippines, New Caledonia, Micronesia, and Polynesia; but vast areas still remained unexplored and botanically unknown.

Thus, except for a few references to Bornean plants in the early work of Rumphius, nothing appears in botanical literature appertaining to the very rich flora of that great island until 1839. New Guinea and its neighboring islands botanically remained practically a *terra incognita* until the last two decades of the nineteenth century, although there are scattered references to New Guinea plants here and there in earlier literature. Even in 1900, on the basis of all the published literature and botanical collections available to them, Schumann and Lauterbach were able to record only about 1,560 species of flowering plants and ferns for all of German New Guinea and the adjacent islands under German control. The area covered embraced also that part of Micronesia that later became the Japanese-mandated territory, covering the Caroline, Marianas, and Marshall Islands. This list includes several hundred introduced and naturalized weeds as well as the introduced and cultivated plants. The number of species now known to occur in the regions covered by the Schumann and Lauterbach work has been vastly increased in the past four decades.

For all of Polynesia and Micronesia, botanical history commences with the year 1773, when the collections made on Cook's first voyage became available to botanists working in England. In the latter part of that century and during the first half of the next century, many contributions appeared in the reports of various exploring expeditions sent out by the British, Spanish, French, Russian, and American governments; but no reasonably complete descriptive flora for any single island appeared for the entire Pacific basin until Seemann's *Flora Vitiensis* was issued in 1865–73, including what was then known about the flora of Fiji.

RESUMPTION OF INTEREST

At about the middle of the last century, local botanical work in British Malaya and in the Netherlands East Indies tended to become static, and after the work of the above-mentioned pioneer botanists appeared, relatively little was accomplished until the last two decades of the century. German botanists, following the establishment of German control over a part of New Guinea and certain neighboring islands, became active in the latter part of the nineteenth century and very greatly increased their activities in the present century. At about the same time, Australian botanists commenced to make known the plants of British New Guinea. As interest in Papuan botany was developed by the Germans, so with the American occupation of the Philippines, beginning in 1902, and following the Japanese control of Micronesia, much activity is recorded in the botanical field covering these areas, quite as the French exploited the flora of New Caledonia and of Tahiti after acquiring control of those islands. Within the present century very much has been accomplished by the Netherlands and British botanists on the floras of the Malaysian region from the Malay Peninsula and Sumatra to New Guinea.

WHAT REMAINS TO BE DONE

While many data have been published in the past two centuries on the vegetation at least of the better known and longer settled islands, a vast amount of work still remains to be accomplished. In Sumatra, Borneo, Celebes, the Moluccas, New Guinea, Bismarck Archipelago, Solomon Islands, New Hebrides, and other islands and island groups, vast areas have never been even visited by a botanist or a botanical collector. Until much more field work has been accomplished and until large comprehensive collections of botanical material are assembled, no one will be in a position to prepare even a reasonably complete treatise covering the plants of many parts of the region under discussion.

Naturally, in the past, the coastal regions and the easily accessible regions at lower altitudes have, in general, been those that have first received attention from explorers and collectors. Thus, from a purely practical standpoint, the vegetation typical of the settled areas, the open grass-lands, the secondary forests, and the coastal districts of the entire region covered by this work and for all or most parts of such regions as Polynesia, Micronesia, the Philippines, the Malay Peninsula, and Java, are now well known. When, however, one considers the primary forests at low, medium, and higher altitudes in other regions than those above noted, the situation is very different, for in these primary forests a very high percentage of all the species found on any single island are of peculiarly restricted geographic distribution. The higher mountains present special problems, for on these the flora is invariably very rich in both genera and species, and local endemism is everywhere highly developed. It will thus be the more inaccessible inland forested regions, and especially the mountains, that will yield the greatest harvest in previously unknown and unnamed species. This fact is emphasized elsewhere in this volume.

BIBLIOGRAPHIC NOTES

Bibliographically, the situation is highly complicated, because of the very large number of technical papers that have been published appertaining to the vast region here considered. Some idea of the number of published items can be gained by the following statement regarding areas where relatively recent comprehensive lists of papers, some short, some extensive, have been issued.

Thus for eastern Asia, covering all of China from Tibet to Yunnan eastward to Manchuria, Korea, Japan, and Formosa, the Merrill-Walker Bibliography (1938) contains in excess of 21,000 author-entries. The bibliography of Philippine botany (1926) contains about 1,700 items. The list of published papers that appertain to Borneo (1915) contains 450 entries. The Polynesian bibliography (1937) lists about 2,600 items, and its as yet unpublished supplement contains about 1,500 more. A similar list for the Malay Archipelago as a whole would be very extensive, as would lists limited to individual islands or island groups, such as Java, Celebes, Sumatra, and New Guinea, and its neighboring islands. While for the archipelago as a whole and for Java one would have to search the European literature issued since the early part of the sixteenth century, for New Guinea and its associated islands most entries would be to literature published within the present century, and in the last two decades of the preceding one, as above noted.

Bibliography is a necessary tool to the working botanist, but is, on the whole, a dry and uninteresting subject. In the entries that follow, no attempt has been made even to list the older titles. Rather, the entries have been restricted to the more recent and more or less standard volumes with some references to papers that originally appeared in periodical literature. Some of these are out of print, or because of their place of publication are not now available. They can, however, be consulted in various institutional libraries. It is believed that some of these references will be of interest to individuals who may be located in the regions covered. Aside from a few general and bibliographic works, the items have been arranged by islands and island groups.

SELECTED BIBLIOGRAPHY

For more detailed information appertaining to many of the regions covered by this volume, the following publications may be consulted. Basic publications on certain regions not discussed in this volume have been added, such as Australia, New Zealand, New Caledonia, China, Indo-China, and Hainan.

BIBLIOGRAPHIC

Blake, S. F. & Atwood, A. C. Geographical Guide to the Floras of the World. An Annotated List with Special Reference to Useful Plants and Common Plant Names. Part I. Africa, Australia, North America, and Islands of the Atlantic, Pacific, and Indian Oceans. U. S. Dept. Agr. Miscel. Publ. 401: 1–336. 1942.

Lam, H. J. Materials Towards a Study of the Flora of the Island of New Guinea. Blumea 1: 115–159. 1934.

Partly bibliographic. The more important published botanical papers appertaining to New Guinea are listed pp. 147–159

Merrill, E. D. Polynesian Botanical Bibliography 1773–1935. Bishop Mus. Bull. 144: 1–194. 1937.

Covers the islands of the Pacific from the Hawaiian and Juan Fernandez Islands to the Marianas and Caroline Islands, southward to New Caledonia. Includes about 2,600 author-entries. The unpublished Supplement contains about 1,500 additional ones.

Merrill, E. D. & Walker, E. H. Bibliography of Eastern Asiatic Botany. i–xlii, 1–719. 1938. Arnold Arboretum, Jamaica Plain, Mass.

Covers all of Japan and eastern Asia inland to Tibet and southward to the Indo-China border. Contains in excess of 21,000 author-entries.

SELECTED BIBLIOGRAPHY

GENERAL

Bentham, G. & Hooker, J. D. Genera plantarum ad exemplaria imprimis in herbariis Kewensibus servata definita, 3 vols. 1862–83. London.
  Treats the families and genera of flowering plants for the entire world.

Engler, A. & Prantl, K. Die natürlichen Pflanzenfamilien. Teil 1–4, illus., and Supplements (Nachträge), 1887–1915. Leipzig. Ed. 2, 1924–1940.
  Treats the families and genera of all groups of plants for the entire world. About 20 volumes of the second edition published to date.

Johnston, I. M. The Preparation of Botanical Specimens for the Herbarium. 1–33, illus., 1939. Arnold Arboretum, Jamaica Plain, Mass.

ECONOMIC BOTANY

Brown, W. H. Minor Products of Philippine Forests, 3 vols., illus. Philippine Bureau of Forestry Bull. 22. 1920–21.
—— Useful Plants of the Philippines. 1: 1–590. fig. 1–253. 1941. Manila.

Burkill, I. H. A Dictionary of the Economic Products of the Malay Peninsula, 2 vols. 1935. London.

Heyne, K. De nuttige planten van Nederlandsch-Indië, 4 vols. 1913–16; ed. 2, 3 vols. 1927. Batavia.

Ochse, J. J. Vegetables of the Dutch East Indies. i–xxxvi, 1–1005. fig. 1–463. 1931. Buitenzorg.
—— Fruits and Fruticulture in the Dutch East Indies. i–xv, 1–180. pl. 1–57. 1931. Batavia.

SPECIAL BOTANICAL PUBLICATIONS ON PARTICULAR ISLANDS OR REGIONS

Aleutian Islands

Hultén, E. Flora of the Aleutian Islands and Westernmost Alaska Peninsula with Notes on the Flora of the Commander Islands. 1–397. 1937. Stockholm.

Collins, H. B., Clark, A. H. & Walker, E. H. The Aleutian Islands, Their People and Natural History. 1–129, illus., 1945.

Smithsonian Institution, Washington. Includes a brief summary of the vegetation with a key to the identification of selected species by E. H. Walker, pp. 63-129.

Kamchatka

Hultén, E. Flora of Kamtchatka and the Adjacent Islands, 2 vols., illus., 1926-30. Stockholm.

Komarov, V. L. Floræ Peninsulæ Kamtchatka, 3 vols., illus., 1927-30. Leningrad.

Kurile Islands

Kudo, Y. Flora of the Island of Paramushir. Jour. Col. Agr. Hokkaido Univ. 11 : 23-183, illus. 1922.

Miyabe, K. The Flora of the Kurile Islands. Mem. Boston Soc. Nat. Hist. 4: 203-275. 1890.

Ohwi, J. Flora Shikotanensis [Flora of Shikotan Island]. Act. Phytotax. Geobot. 1 : 34-55, 111-131. 1932; 3 : 263-287. 1933.

Tatewaki, M. Vascular Plants of the Middle Kuriles. Bull. Biogeogr. Soc. Japan 4: 257-334, illus. 1934.

—— The Phytogeography of the Middle Kuriles. Jour. Fac. Agr. Hokkaido Univ. 29: 151-363, illus. 1933.

Sakhalin

Sugawara, S. Illustrated Flora of Saghalien, 4 vols. 1-1957, pl. 1-892. 1937-40. Tokyo.

Japan

Makino, T. Futsu shokobutsu dzufu (Illustrations of Common Plants), 5 vols., many illustrations in color. 1912. Tokyo. (Japanese text).

—— **& Nemoto, K.** Nippon shokobutsu soran (Flora of Japan), ed. 2, 1-1036. 1931; Soran hoi (Supplement) by K. Nemoto, 1-1436. 1936. Tokyo. (Japanese text).

—— **& Tanaka, K.** A Manual of the Flora of Nippon. 1-862, fig. 1-212. 1927. Tokyo. (Japanese text).

Murakoshi, M. Nai-gwai shokobutsu genshoku dai-dzukan (Illustrated Encyclopedia of Botany). 13 vols. 1935. Tokyo. (Japanese text).

Includes 375 plates with 2044 figures of plants.

—— Genshoku dzusetsu shokobutsu daijiten (A Comprehensive Botanical Dictionary with Illustrations in Color). 1-472, indices, 236 plates. 1938. Tokyo. (Japanese text).

Terasaki, T. Nippon shokobutsu dzufu (Illustrations of Japanese Plants). fig. 1–2100. 1933; Zoku hen (Supplement), fig. 2101–4000. 1938. Tokyo. (Japanese text).

China

For detailed information regarding all known publications on the botany of this area, including also Japan, the Riu Kiu Islands, and Formosa, as well as southern and eastern Siberia, Manchuria, and Korea, see **Merrill, E. D. & Walker, E. H.** Bibliography of Eastern Asiatic Botany. i–xlii, 1–719. 1938. Arnold Arboretum, Jamaica Plain, Mass.

Indo-China

Lecomte, H. Flore générale de l'Indo-Chine, 7 vols. and Supplement, illus. 1907–1939. Paris.

Hainan

Merrill, E. D. An Enumeration of Hainan Plants. Lingnan Sci. Jour. 5: 1–186. 1927.
—— A Working List of Hainan Plants. Unpublished manuscript. 1944.
Tanaka, T. & Odashima, K. A Census of Hainan Plants. Jour. Soc. Trop. Agr. 10: 357–402. 1938.
Odashima, K. & Tanaka, T. Supplement to the Census of Hainan Plants. Jour. Soc. Trop. Agr. 12: 193–204. 1940.

Formosa

Hayata, B. Icones Plantarum Formosanarum, 10 vols., illus., 1911–21. Tokyo. Contains numerous plates and figures of plants.

Galápagos Islands

Robinson, B. L. Flora of the Galapagos Islands. Proc. Amer. Acad. Arts & Sci. 38: 77–269, illus., 1902 (Contributions from the Gray Herbarium No. 24). Bibliography pp. 80–82.
Stewart, A. A Botanical Survey of the Galapagos Islands. Proc. Calif. Acad. Sci. IV. 1: 7–252, illus., 1911.
Bibliography pp. 246–248.

Hawaii

Degener, O. Flora Hawaiiensis, or New Illustrated Flora of the Hawaiian Islands, 4 vols. 1932–40. Honolulu.

Illustrations and complete descriptions of 400 native and introduced species.

Hillebrand, W. Flora of the Hawaiian Islands; a Description of Their Phanerogams and Vascular Cryptogams, i–xcvi, 1–673. 1888. London.

Rock, J. F. The Indigenous Trees of the Hawaiian Islands. 1–518, illus., 1913. Honolulu.

—— The Ornamental Trees of Hawaii. 1–210, illus., 1917. Honolulu.

Micronesia

Diels, L. Beiträge zur Flora von Mikronesien und Polynesien. I–V. Bot. Jahrb. 52: 1–18. 1914 to 69: 395–400. 1938.

Kanehira, R. Flora Micronesica, 1–8, 1–468, 1–37, illus., 1933. Tokyo. (Japanese text).

Johnston and Wake Islands

Christophersen, E. Vascular Plants of Johnston and Wake Islands. Occ. Pap. Bishop Mus. 9(13): 1–20, illus., 1931.

Guam

Merrill, E. D. An Enumeration of the Plants of Guam. Philippine Jour. Sci. 9: Bot. 17–155. 1914.

Safford, W. E. The Useful Plants of Guam. Contributions U. S. Nat. Herb. 9: 1–416, illus., 1905.

Tahiti and Neighboring Islands

Drake del Castillo, E. Flora de la Polynésie française. Description des plantes vasculaires qui crossent spontanément ou qui sont généralement cultivées aux îles de la Société, Marquesas, Pomotou, Gambier et Wallis. i–xxiv, 1–352. 1892. Paris.

Palmyra

Rock, J. F. Palmyra Island, With a Description of Its Flora. College of Hawaii Bull. 4: 1–53, illus., 1916.

Christmas, Jarvis, Fanning and Neighboring Islands

Christophersen, E. Vegetation of Pacific Equatorial Islands. Bishop Mus. Bull. 44: 1–79, illus., 1927.

Marquesas and Neighboring Islands

Brown, F. H. B. Flora of Southeastern Polynesia. Bishop Mus. Bull. 84: 1–194, illus., 1931; 89: 1–123, illus., 1931; 130: 1–386, illus., 1935.

Samoa

Christophersen, E. Flowering Plants of Samoa. Bishop Mus. Bull. 128: 1–221, illus., 1935; 154: 1–77, illus., 1938.

Reinecke, F. Die Flora der Samoa-Inseln. Bot. Jahrb. 23: 237–368. 1896; 25: 578–708. 1898.

Setchel, W. A. American Samoa. Carnegie Inst. Publ. 341: 1–275, illus., 1924.

Fiji

Gibbs, L. S. A Contribution to the Montane Flora of Fiji. Journ. Linn. Soc. Bot. 39: 130–212, illus., 1909.

Gillespie, J. W. New Plants from Fiji. Bishop Mus. Bull. 74: 1–99, illus., 1930; 83: 1–72, illus., 1931; 91: 1–43, illus., 1932.

Mead, J. P. The Forests of the Colony of Fiji. Fiji Legislative Council Paper 4: 1–47, 2 maps, 1928. Suva.

Appertains to forest resources and commercial timber trees.

Seemann, B. Flora Vitiensis; a Description of the Plants of the Viti or Fiji Islands. i–xxxii, 1–453, illus., 1865–73. London.

Smith, A. C. Fijian Plant Studies. Bishop Mus. Bull. 141: 1–166, illus., 1936; (II) Botanical Results of the 1940–41 Cruise of the "Cheng Ho." Sargentia 1: 1–148. 1942; (III) New and Noteworthy Flowering Plants from Fiji. Bull. Torrey Bot. Club. 70: 533–549. 1943.

Tonga

Burkill, I. H. The Flora of Vavao, one of the Tonga Islands, Jour. Linn. Soc. Bot. 35: 20–65, illus., 1901.

Hemsley, W. B. The Flora of the Tonga or Friendly Islands. Jour. Linn. Soc. Bot. 30: 158–217, illus., 1894.

Rarotonga

Cheeseman, T. F. The Flora of Rarotonga, the Chief Island of the Cook Group. Trans. Linn. Soc. II, Bot. 6: 261–313, illus., 1903.

Wilder, G. P. Flora Rarotonga. Bishop Mus. Bull. 86: 1–111, illus., 1931.

Makatea

Wilder, G. P. The Flora of Makatea. Bishop Mus. Bull. 120: 1–49, illus., 1934.

Niue

Yuncker, T. G. The Flora of Niue Island. Bishop Mus. Bull. 178: 1–126, illus., 1943.

Funafuti

Maiden, J. H. The Botany of Funafuti, Ellice Group. Proc. Linn. Soc. New South Wales 29: 539–556. 1904.

New Hebrides

Guillaumin, A. Contributions to the Flora of the New Hebrides. Plants collected by S. F. Kajewski in 1928 and 1929. Jour. Arnold Arb. 12: 221–264. 1931; 13: 1–29, 81–126. 1932.

For other data see Merrill, E. D., Polynesian Botanical Bibliography 1773–1935, Bishop Mus. Bull. 144: 91–92.

New Caledonia

Guillaumin, A. Matériaux pour la flore de la Nouvelle-Calédonie 1(1914)–55(1939).

Published in various periodicals.

—— Contribution à la flore de la Nouvelle-Calédonie 1 (1911)–70 (1939).

Published in various periodicals. For details regarding this entry and the preceding one see Merrill, E. D., Polynesian Botanical Bibliography 1773–1935, Bishop Mus. Bull. 144: 89–91. 1937.

New Guinea and Neighboring Islands

Gibbs, L. S. A Contribution to the Phytogeography and Flora of the Arfak Mountains. 1–222, illus., 1917. London.

Lam, H. J. Fragmenta Papuana I–VII. Natuurk. Tijdschr. Nederl.-Ind. 87: 111–130, illus., 1927 to 89: 291–388, illus., 1929. English translation by L. M. Perry as: Fragmenta Papuana [Observations of a Naturalist in Netherlands New Guinea]. Sargentia 5: 1–196. fig. 1–32. 1945. Arnold Arboretum, Jamaica Plain, Mass.

General notes on the country and vegetation made on his trip up the Mamberamo River to the Central Mountain range of Netherlands New Guinea in 1920.

Lane-Poole, C. E. The Forest Resources of the Territories of Papua and New Guinea. Australian Parliamentary Paper 1–209. 1925. Melbourne.

Lauterbach, K. Beiträge zur Flora von Papuasien I–XXVI. Bot. Jahrb. 49: 1–169, illus., 1912 to 72: 155–269, illus., 1942.
Numbers 23 to 26 edited by L. Diels.

Lorentz, H. A. (editor). Nova Guinea. Résultats de l'expédition scientifique néerlandaise à la Nouvelle-Guinée—Botanique. Vols. 8, 12, 14, 18, illus., 1909–1936. Leiden.

Merrill, E. D. & Perry, L. M. Plantæ Papuanæ Archboldianæ I–XV. Jour. Arnold Arb. 20: 324–345 (1939) to 26: 1–36. 1945.

Ridley, H. N. Report on the Botany of the Wollaston Expedition to Dutch New Guinea, 1912–13. Trans. Linn. Soc. II. Bot. 9: 1–269, illus., 1916. London.

Schumann, K. & Lauterbach, K. Die Flora der deutschen Schutzgebiete in der Südsee. i–xvi, 1–613, illus., 1900; Nachtrage 1–446, illus., 1905. Leipzig.

Smith, A. C. Studies on Papuasian Plants I–VI. Jour. Arnold Arb. 22: 60–80. 1941 to 25: 222–270. (1944).

Philippine Islands

Brown, W. H. Vegetation of Philippine Mountains. The Relation Between the Environment and Physical Types at Different Altitudes. 1–434. pl. 1–41. 1919. Bureau of Science. Manila.

Merrill, E. D. A Flora of Manila. 1–490. 1912. Bureau of Science, Manila.

A general descriptive flora which includes most of the species found in settled areas at low altitudes for the entire Malaysian region.

—— An Enumeration of Philippine Flowering Plants, 4 vols., 1923–1926. Bibliography 4: 155–239. Bureau of Science, Manila.

Borneo

Merrill, E. D. A Bibliographic Enumeration of Bornean Plants. 1–637. 1921. Singapore.

Malay Peninsula

Corner, E. J. H. Wayside Trees of Malaya, 2 vols., illus., 1940. Singapore.

Ridley, H. N. The Flora of the Malay Peninsula, 5 vols., illus., 1922–25. London.
A general descriptive flora.

Java

Koorders, S. H. Exkursionsflora von Java, umfassend die Blütenpflanzen mit Besonderer Berücksichtigung der im Hochgebirge Wildwachsenden Arten, 3 vols., atlas, fig. 1–1313. 1911–37. Jena.
—— **& Valeton, T.** Bijdrage tot de kennis der boomsoorten van Java. 1(1894)–13(1914). Batavia.
—— Atlas der Baumarten von Java. 1(1913)–4(1918). Batavia.
Companion volumes to the preceding item, containing 800 illustrations of tree species.

New Zealand

Cheeseman, T. F. Manual of the New Zealand Flora, ed. 2, i–xi, 1–1163. 1925. Wellington.

Australia

Bailey, F. M. The Queensland Flora, 6 vols., 1899–1902. Brisbane.
Bentham, G. & Mueller, F. von. Flora Australiensis, 7 vols., 1863–1878. London.
Black, J. M. Flora of South Australia. 1–745. 1922–29. Adelaide.
Ewart, A. J. & Davies, O. B. The Flora of the Northern Territory. viii, 1–387, illus., 1917. Melbourne.
Ewart, A. J. Flora of Victoria, 1–1257. 1930. Melbourne.
Moore, C. & Betche, E. Handbook of the Flora of New South Wales. i–xxix, 1–582. 1893. Sydney.

16.
Simple Directions for Preparing Botanical Specimens

The technique of preparing dried specimens of plants for identification purposes is very simple. The objective is to extract the moisture in the shortest possible time, in order to avoid undue discoloration, and at the same time to keep the specimens flat and under moderate pressure. Presses containing selected specimens to be dried may be kept in a warm sunny place during the daytime, but should be placed under shelter at night. In very damp or rainy weather it is best to use artificial heat, the presses

Fig. 256. A simple botanical press.

being kept in proximity to a fire or any heating or cooking unit. Artificial heat under any circumstances will greatly expedite the drying process, but care should be taken not to apply too much heat for this would "cook" the specimens. Press frames (fig. 256), about 18 by 12 inches in size, may be made from strips of bamboo or of material taken from ordinary packing cases. If this size be used, then the prepared material will be adapted to the standard size of herbarium sheets used in all American institutions. Pressure may be applied by stout cords, one tied near each of the four corners of the lower press frame; these cords may be tightened from time to time as the contents of the press shrink in drying, and when driers are changed. However, a press is not a real essential, as stacks of drying plants may be arranged between a couple of boards, pressure being supplied by a rock weighing about 40 pounds.

Specimens should be thoroughly dry before being removed from the presses; otherwise they will mold and decay. The packages of prepared specimens should be kept in a dry place. From time to time they may be packed with sufficient protection to prevent breaking in transit, and shipped by parcel post.

IMPROVISED DRIERS

It is essential that the paper used for drying be changed at least once a day, and preferably more frequently, the damp papers being replaced by fresh, dry ones. Frequent changes will expedite drying. Any absorbent material such as blotting paper may be used as driers, or even old newspapers and pages from discarded "pulp" magazines; several sheets are placed between specimens that are being dried. Excellent driers may be made from the corrugated paper from discarded packing cases, provided such cases have not been treated with some moisture-repellent material. Damp driers should be thoroughly dried, in the sun or near a fire, before they are again used. The standard size of driers is about 18 by 12 inches, but smaller sizes may be used. Glazed paper should not be used, but such paper is excellent for storing the specimens when they are once dried. It is hence quite possible to improvise in the field all the material essential to the actual preparation of botanical specimens. Some plants such as grasses and sedges, and small herbs and ferns, will dry very quickly. Others will take a longer time. Very fleshy plants dry slowly; this is particularly true of the leaves of epiphytic orchids.

SELECTION OF SPECIMENS

Some care should be used in the selection of specimens, for the portions of a plant selected for drying should be reasonably representative. For small herbs, grasses, sedges, and ferns, take the whole plant. From a tree, shrub, or coarse vine, take a portion about 16 inches long. Tall grasses and sedges may be reduced by being bent back and forth, and to keep such often wiry material in place in the press, slip a slit paper over the angles. With very tall grasses, take only parts of the plant that are sufficient to show its general characters. In the case of large ferns, portions of the frond, such as all of the branches or leaves on one side of the stem, may be removed. Whenever possible, try to select specimens in flower or in fruit, or both. The position of the leaves on the stem or branch is important, and the leaves, even when fairly large, should not be stripped from the stems. Very large leaves may be folded back and forth to make them come within the limits of the material used for driers; when the leaves are very numerous, some of them may be discarded.

NOTES

Notes may be written on the margins of the sheets, or preferably on slips of paper placed in the folder in which the specimens are arranged for drying. *In general, such notes should indicate those characters that the dried specimen will not show,* such as approximate size, if the plant be large; habitat (that is, whether on the seashore, in thickets, in forests, in swamps, in open grasslands, etc.); collector's name; locality, if this be permissible (in any case, the locality should be recorded and these data sent later when conditions permit); approximate altitude; color of the flower and fruit; local names and economic uses, if the collector be interested in the latter. If milky sap be present, this should be noted. It is admitted that while all of these data are desirable, it is not always possible to record them all; what can be supplied will be of very great value to whoever has the task of identifying and reporting on the material. Small fruits and detached flowers may be placed in packets with the specimens, but whenever possible the actual specimens should show the attachment of both flowers and fruit. Very large fruits may be dried separately or, in some cases, sliced and dried. Each collector should adopt a serial number system, so that the names may be returned to him by number.

COMMON *VERSUS* RARE SPECIES

Most of the plant species that will be observed in the low-altitude settled areas, along the seashores, and even in the secondary forests of any part of the Old World tropics, and all of the cultivated plants, will represent rather widely distributed and well-known ones. Nevertheless, it is desirable from all standpoints that such species be collected, for we need much additional information regarding the occurrence and geographical distribution of all species. Generally speaking, American botanical institutions are weak in their reference collections from the Old World tropics.

If one be more interested in securing novelties, then collecting should largely be confined to the primary forests, for a very high percentage of all species occurring therein have very restricted ranges. This is particularly true of altitudes at and above approximately 3,000 feet, for the higher mountains everywhere in the tropics support rich floras, and it is at and above such an altitude that the vast majority of the orchids, ferns, and other herbaceous species will be found. The mountain floras are everywhere very different from the low-altitude ones.

Vast areas in New Guinea and its adjacent islands, large parts of such islands as Borneo, Celebes, and Sumatra, and many of the neighboring islands have never been explored by a botanist or a botanical collector. General collections in such regions always yield a certain percentage of unnamed and undescribed species. No part of the entire region is exhaustively explored, except parts of Java, the Philippines, and the Malay Peninsula, and even in the most intensively explored parts of these regions new species constantly appear in current collections. On the continent itself, this same condition of totally inadequate botanical exploration applies to most of China, Indo-China, Siam, and Burma.

REGIONS OF SPECIAL INTEREST

Within the present century, many collections have been made in previously unexplored parts of Malaysia where as high as

10 per cent of all species collected proved, on study, to be unnamed and undescribed. Thus if one seeks for novelties, these little-known regions are among the most promising in the entire world. One does not have to be a trained botanist to collect undescribed species, for it is inevitable that many of these will appear as general collections are made. The best advice that one can give is that, circumstances permitting, the collector should attempt to prepare specimens representing all the different species that he may observe in flower or in fruit. However, it is possible for the trained and experienced botanist to name sterile material, at least to the genus. Not infrequently, because of potential economic uses of certain species, the collector may desire to secure identifications; in such cases, specimens without flowers or fruits may be prepared.

IDENTIFICATIONS

If the individual, no matter where located in the Orient, should care to secure the names and some idea of the relationships of plants that may interest him, packages of specimens may be addressed to the Director, Arnold Arboretum, Jamaica Plain, Massachusetts, U.S.A. This is the only American botanical institution whose staff members are actually specializing on the study of plants from all parts of the Old World tropics, particularly Polynesia and Micronesia, and the entire Southwest Pacific region, the Malay Archipelago, and all parts of tropical and subtropical Asia. Normally, lists of identifications can be sent by return mail, although a certain percentage of determinations will be to the genus only. Specimens should be numbered serially, so that the names may be returned by number. Most important contributions may thus be made to our knowledge of Old World botany.

SHIPMENT OF MATERIAL

Packages should be tied tightly, wrapped well in several thicknesses of paper, and again tied. Cardboard, corrugated paper, thin boards, or even woven sheets of thinly split bamboo may be used to protect the material against breakage in transit. Packages should be marked "dried botanical material for scientific study," for such material is not subject to quarantine, *nor is an importation permit required*. Packages so marked normally will come through by parcel post without delay. If one wishes to ship living material to the United States, a permit must be secured from the Bureau of Entomology and Plant Quarantine, U. S. Department of Agriculture, Washington, D.C., for living material is subject to examination and quarantine because of the danger of introduction of noxious insects and plant diseases.

Botanical Arrangement of the Species

DISCUSSED AND ILLUSTRATED, WITH PAGE AND FIGURE REFERENCES.

Pteridophyta (ferns):
  Acrostichum aureum Linn., 52, 108, 181, fig. 66A.
  Angiopteris evecta (Forst.) Hoffm., 110.
  Antrophyum plantagineum (Cav.) Kaulf., 110, fig. 103.
  Asplenium nidus Linn., 55, 106.
  Athyrium esculentum (Retz.) Copel., 181.
  Cheiropleuria bicuspis (Blume) Presl, 110, fig. 107.
  Cyclophorus lanceolatus (Linn.) Alst., 110, fig. 104.
  Dipteris conjugata Reinw., 110, fig. 108.
  Drymoglossum carnosum (Wall.) J. Sm., 110, fig. 111.
  Drynaria quercifolia (Linn.) J. Sm., 52, 108, fig. 110.
  Helminthostachys zeylanica (Linn.) Hook., 108, fig. 106.
  Hymenolepis spicata (Linn.f.) Presl, 107, fig. 101.
  Lecanopteris carnosa (Reinw.) Blume, 99, 110, fig. 95.
  Lygodium circinnatum (Burm.f.) Sw., 108, fig. 105.
  Ophioglossum pendulum Linn., 108, fig. 102.
  Photinopteris speciosa Blume, 110, fig. 109.
  Polypodium heracleum Kunze, 108.
    sinuosum Wall., 55, 99, 110, fig. 94.
  Stenochlaena palustris, (Burm.f.) Bedd., 53, 108, 182, fig. 66B.
Cycadaceae:
  Cycas circinalis Linn., 30, 187, fig. 12.
Pinaceae:
  Agathis alba (Lam.) Foxw., 82.
Gnetaceae:
  Gnetum gnemon Linn., 61, 187, 222, fig. 69.
Pandanaceae:
  Pandanus brosimos Merr. & Perr., 190.
    conoideus Lam., 190.
    julianettii, Mart., 190.
    tectorius Sol., 30, 54, 190, fig. 11.
Gramineae:
  Andropogon sorghum (Linn.) Brot., 147.
  Coix lachryma jobi Linn., 147, 184.
  Cynodon dactylon (Linn.) Pers., 123.
  Eleusine corocana (Linn.) Gaertn., 147.
    indica (Linn.) Gaertn., 123.

Imperata cylindrica (Linn.) Beauv., 65.
Ischaemum muticum Linn., 34.
Oryza sativa Linn., 147.
Saccharum officinarum Linn., 148, 166.
  spontaneum Linn., 65, 184.
Setaria italica (Linn.) Beauv., 147, 228.
Spinifex littoreus (Burm.f.) Merr., 34, 37, fig. 13.
Thuarea involuta (Forst.) Roem. & Schult., 34.
Thysanolaena maxima (Roxb.) O. Kuntze, 65.

Cyperaceae:
  Remirea maritima Aubl., 39.

Palmae:
  Arenga pinnata (Wurmb) Merr., 163, 164, 182.
  Areca catechu Linn., 162, 164.
  Borassus flabellifer Linn., 163.
  Cocos nucifera Linn., 34, 162, 163, 166, 182.
  Corypha elata Roxb., 163.
  Elaeis guineensis Jacq., 162, 166.
  Licuala spinosa Wurmb, 35.
  Metroxylon sagu Rottb., 164.
  Nipa fruticans Wurmb, 34, 54, 164, 182.
  Oncosperma filamentosum Blume, 164.
  Roystonea regia (H.B.K.) Cook, 164.

Araceae:
  Alocasia macrorrhiza (Linn.) Schott, 104, fig. 98.
  Amorphophallus campanulatus (Roxb.) Blume, 103, fig. 100.
    titanum Becc., 103, fig. 99.
  Caladium bicolor Vent., 165.
  Colocasia esculenta (Linn.) Schott, 149, fig. 190.
  Cyrtosperma merkusii (Hassk.) Schott, 104.
  Xanthosoma violaceum Schott, 149.

Flagellariaceae:
  Flagellaria indica Linn., 35.

Bromeliaceae:
  Ananas comosus (Linn.) Merr., 151.

Commelinaceae:
  Commelina benghalensis Linn., 125, fig. 114.
    diffusa Burm.f., 125, fig. 117.
  Cyanotis cristata (Linn.) Roem. & Schult., 125, fig. 116.
  Rhoeo discolor (L'Hérit.) Hance, 165, fig. 219.

Liliaceae:
  Allium tuberosum Spreng., 228.
  Cordyline fruticosa (Linn.) A. Chev., 159, fig. 220.
  Sansevieria roxburghiana Schult.f., 159.

BOTANICAL ARRANGEMENT OF THE SPECIES 265

Amaryllidaceae:
  Crinum asiaticum Linn., 35, fig. 14.
  Polianthes tuberosa Linn., 165.
Taccaceae:
  Tacca pinnatifida Forst., 35, 185, fig. 15.
Dioscoreaceae:
  Dioscorea alata Linn., 149, 185, 224, fig. 181.
    esculenta (Lour.) Burk., 61, 149, 185, fig. 67.
    hispida Dennst., 61, 185, fig. 68.
Musaceae:
  Musa textilis Née, 166.
    uranoscopos Linn., 110.
  Ravenala madagascariensis J. F. Gmel., 156, fig. 221.
Zingiberaceae:
  Hedychium coronarium Koen., 228.
  Zingiber officinale Rosc., 166.
Orchidaceae:
  Grammatophyllum speciosum Blume, 113.
  Phalaenopsis amabilis (Linn.) Blume, 110, fig. 9C.
Casuarinaceae:
  Casuarina equisetifolia Linn., 28, 29, fig. 18.
Piperaceae:
  Piper betle Linn., 164.
    cubeba Linn., 166.
    myrmecophilum C. DC., 98.
    nigrum Linn., 166.
Ulmaceae:
  Trema orientalis (Linn.) Blume, 60, fig. 71.
Moraceae:
  Antiaris toxicaria (Pers.) Lesch., 1.
  Artocarpus altilis (Park.) Fosb., 153, 188, fig. 193.
    champeden (Lour.) Spreng., 95, 155.
    heterophylla Lam., 95, 228, fig. 193.
    odoratissima Blanco, 155.
    rotunda (Houtt.) Merr., 155.
  Ficus benjamina Linn., 158.
    elastica Linn., 158.
    mirabilis Merr., 96.
    religiosa Linn., 158.
Urticaceae:
  Laportea, 3, fig. 3.
  Pilea microphylla (Linn.) Liebm., 126.
  Pouzolzia zeylanica (Linn.) Benn., 126.

Olacaceae:
: Ximenia americana Linn., 31, 39, fig. 22.

Rafflesiaceae:
: Rafflesia arnoldii R. Br., 101, fig. 86.
: schadenbergiana Gopp., 101.

Polygonaceae:
: Antigonon leptopus Hook. & Arn., 161, fig. 222.
: Muehlenbeckia platyclada Meisn., 160, fig. 224.

Chenopodiaceae:
: Chenopodium ambrosioides Linn., 126, 229, fig. 124.

Amaranthaceae:
: Achyranthes aspera Linn., 126, fig. 115.
: Alternanthera sessilis (Linn.) R. Br., 126, fig. 112.
: versicolor Regel, 165.
: Amaranthus spinosus Linn., 126, fig. 120.
: tricolor Linn., 165.
: viridis Linn., 126, fig. 118.
: Celosia argentea Linn., 126, fig. 113.
: cristata Linn., 165.

Nyctaginaceae:
: Boerhavia diffusa Linn., 35, 179, 192, fig. 17.
: Bougainvillea spectabilis Willd., 161, fig. 223.

Aizoaceae:
: Sesuvium portulacastrum Linn., 36, 39, 179, 191, fig. 20.

Portulacaceae:
: Portulaca oleracea Linn., 36, 126, 179, 191, fig. 122.

Magnoliaceae:
: Michelia alba DC., 157.
: champaca Linn., 157, 228, fig. 225.

Annonaceae:
: Annona muricata Linn., 152, 230, fig. 194.
: reticulata Linn., 152, fig. 195.
: squamosa Linn., 152, fig. 192.
: Cananga odorata (Lam.) Hook.f. & Th., 156, fig. 228.

Myristicaceae:
: Myristica fragrans Houtt., 166.
: myrmecophila Becc., 97.

Lauraceae:
: Cassytha filiformis Linn., 36, 39, fig. 24.
: Cinnamomum zeylanicum Blume, 166.
: Tetranthera salicifolia Becc., 100, fig. 97K.

Hernandiaceae:
: Gyrocarpus americanus Jacq., 82.
: Hernandia ovigera Linn., 29, fig. 16.
: peltata Meisn., 29, fig. 16.

BOTANICAL ARRANGEMENT OF THE SPECIES 267

Papaveraceae:
  Argemone mexicana Linn., 126, fig. 123.
Cruciferae:
  Brassica pekinensis (Lour.) Rupr., 228.
Capparidaceae:
  Gynandropsis gynandra (Linn.) Briq., 126, fig. 119.
  Polanisia icosandra (Linn.) Wight & Arn., 126, fig. 121.
Moringaceae:
  Moringa oleifera Lam., 157, 191.
Nepenthaceae:
  Nepenthes clipeata Dans., 3, fig. 1B.
    ephippiata Dans., 3, fig. 1C.
    inermis Dans., 3, fig. 1D.
    merrilliana Macf., 3, fig. 1F.
    papuana Dans., 3, fig. 1A.
    treubiana Warb., 3, fig. 1E.
Leguminosae:
  Abrus precatorius Linn., 127, fig. 129.
  Acacia farnesiana (Linn.) Willd., 62, fig. 75.
  Adenanthera pavonina Linn., 157.
  Albizzia lebbeck (Linn.) Benth., 157.
  Bauhinia acuminata Linn., 160, fig. 227.
    monandra Kurz, 160.
    purpurea Linn., 160.
    tomentosa Linn., 160, fig. 226.
  Caesalpinia crista Linn., 39, 53.
    nuga (Linn.) Ait., 53, fig. 58.
    pulcherrima (Linn.) Sw., 159, fig. 229.
  Cajanus cajan (Linn.) Millsp., 148, fig. 197.
  Canavalia maritima (Aubl.) Thouars, 36, 39, fig. 25.
    microcarpa (DC.) Piper, 37, fig. 25B.
  Cassia alata Linn., 126, 229, fig. 127.
    fistula Linn., 157, fig. 231.
    javanica Linn., 157.
    nodosa Ham., 157.
    occidentalis Linn., 126, fig. 125.
    siamea Linn., 157, fig. 232.
    tora Linn., 126, 127, fig. 134.
  Clitoria ternatea Linn., 127, fig. 141.
  Crotalaria incana Linn., 127, fig. 137.
    mucronata Desv., 127, fig. 135.
    quinquefolia Linn., 127, fig. 139.
    retusa Linn., 127, fig. 136.
    verrucosa Linn., 127.
  Cynometra cauliflora Linn., 94, 155, fig. 87.

Delonix regia (Boj.) Raf., 156, fig. 234.
Derris trifoliata Lour., 53, fig. 59.
Desmodium gangeticum (Linn.) DC., 137, fig. 131.
  pulchellum (Linn.) Benth., 137, fig. 130.
  umbellatum (Linn.) DC., 32, 53, fig. 21.
  velutinum (Willd.) DC., 137, fig. 132.
Dolichos lablab Linn., 148, fig. 198.
Entada phaseoloides (Linn.) Merr., 39, 103.
Erythrina variegata Linn., 29, 191, fig. 27.
Flemingia strobilifera (Linn.) R. Br., 137, fig. 134.
Gliricidia sepium (Jacq.) Steud., 229.
Indigofera hirsuta Linn., 136, fig. 126.
  suffruticosa Linn., 136, fig. 128A.
  tinctoria Linn., 136, fig. 128B.
Inocarpus fagiferus (Park.) Fosb., 154, 186, fig. 202.
Leucaena glauca (Linn.) Benth., 62, fig. 138.
Mimosa pudica Linn., 127, fig. 138.
Mucuna sp., 6, fig. 2.
  albertisii Becc., 102.
Pachyrrhizus erosus (Linn.) Urb., 148, 186, fig. 200.
Parkia roxburghii G. Don, 18.
Peltophorum pterocarpum (DC.) Backer, 157.
Phaseolus aureus Roxb., 148.
  calcaratus Roxb., 148.
  lunatus Linn., 148, fig. 201.
Pithecellobium dulce (Roxb.) Benth., 229, 230.
Pongamia pinnata (Linn.) Pierre, 32, fig. 19.
Psophocarpus tetragonolobus (Linn.) DC., 148, fig. 196.
Pterocarpus indicus Willd., 157.
Samanea saman (Jacq.) Merr., 156.
Sesbania grandiflora (Linn.) Pers., 157, 191, fig. 230.
Sophora tomentosa Linn., 31, fig. 28.
Tamarindus indica Linn., 157, fig. 100.
Vigna marina (Burm.) Merr., 36, 39, fig. 26.
  sinensis (Linn.) Savi, 148.

Oxalidaceae:

Averrhoa bilimbi Linn., 153, fig. 203A.
  carambola Linn., 94, 153, fig. 203B.
Oxalis corniculata Linn., 126.
  repens Thunb., 126, fig. 140.

Rutaceae:

Atalantia linearis Blanco, 100, fig. 97I.
Murraya paniculata (Linn.) Jack, 159.

Simarubaceae:
  Suriana maritima Linn., 33, fig. 23.

Burseraceae:
  Canarium commune Linn., 187, fig. 70.
    ovatum Engl., 187.

Meliaceae:
  Lansium domesticum Jack, 94, 154, fig. 204.
  Melia parasitica Osb., 94.
  Sandoricum koetjape (Burm.f.) Merr., 154.
  Xylocarpus granatum Koen., 52, fig. 56.
    moluccensis M. Roem., 31, fig. 29.

Malpighiaceae:
  Tristellateia australasiae A. Rich., 53, fig. 65.

Euphorbiaceae:
  Acalypha hispida Burm.f., 158, fig. 236.
    indica Linn., 127, fig. 146.
    wilkesiana Muell.-Arg., 158, fig. 235.
  Aleurites moluccana Willd., 61, fig. 81.
  Antidesma ghaesembilla Gaertn., 61, fig. 80.
  Baccaurea, racemosa Muell.-Arg., 155.
  Codiaeum variegatum (Linn.) Blume, 159, fig. 237.
  Croton tiglium Linn., 160, 222.
  Endospermum formicarum Becc., 97, fig. 89.
  Euphorbia atoto Forst., 36, fig. 33.
    heterophylla Linn., 127, fig. 142.
    hirta Linn., 127, fig. 143.
    prostrata Ait., 127.
    pulcherrima Willd., 258, fig. 238.
    thymifolia Burm.f., 127.
  Excoecaria agallocha Linn., 5, 31, 54, fig. 31.
  Hevea brasiliensis (H.B.K.) Muell.-Arg., 166.
  Homalanthus populneus (Geisel.) Pax, 61, fig. 74.
  Homonoia riparia Lour., 100, fig. 97G.
  Jatropha curcas Linn., 61, 160, fig. 84.
  Macaranga caladifolia Becc., 97, fig. 88.
    philippensis (Lam.) Muell.-Arg., 61, fig. 76.
    tanarius (Linn.) Muell.-Arg., 61, fig. 72.
  Mallotus ricinoides (Pers.) Muell.-Arg., 61, fig. 77.
  Manihot esculenta Crantz, 149, 166, fig. 205.
  Melanolepis multiglandulosa (Reinw.) Rchb.f. & Zoll., 61, fig. 82.
  Phyllanthus niruri Linn., 127, fig. 144.
    urinaria Linn., 127.
  Ricinus communis Linn., 127, fig. 145.

Anacardiaceae:
- Anacardium occidentale Linn., 151, 189, fig. 206.
- Campnosperma macrophylla Hook.f., 82.
- Mangifera caesia Jack, 151, 188.
    - foetida Lour., 151, 189.
    - indica Linn., 151, fig. 207.
    - lagenifera Griff., 151, 188.
    - odorata Griff., 151, 188.
- Semecarpus cuneifolius Blanco, 4, fig. 4.
    - gigantifolius Vidal, 95.
- Spondias cytherea Sonn., 151.
    - pinnata (Linn.) Kurz, 151.

Sapindaceae:
- Cardiospermum halicacabum Linn., 127, fig. 147.
- Dodonaea viscosa (Linn.) Jacq., 31, fig. 32.
- Litchi chinensis Sonn., 153.
- Nephelium lappaceum Linn., 153, fig. 208A.
    - mutabile Blume, 153, fig. 208B.

Rhamnaceae:
- Colubrina asiatica (Linn.) Brongn., 33, fig. 25.

Tiliaceae:
- Corchorus acutangulus Lam., 128.
    - capsularis Linn., 128.
    - olitorius Linn., 128, fig. 151.
- Triumfetta bartramia Linn., 128, fig. 149.
    - procumbens Forst., 36, fig. 30.
    - repens (Blume) Merr. & Rolfe, 36.
    - semitriloba Jacq., 128.

Malvaceae:
- Abelmoschus moschatus Medic., 128, 228, fig. 155.
- Abutilon indicum (Linn.) Sweet, 128, fig. 148.
- Hibiscus mutabilis Linn., 159, fig. 239.
    - rosa-sinensis Linn., 159, fig. 241.
    - schizopetalus (Mast.) Hook.f., 159, fig. 240.
    - surattensis Linn., 128, fig. 159.
    - tiliaceus Linn., 30, 39, 54, 222, fig. 37.
- Malvastrum coromandelinum (Linn.) Garcke, 128, fig. 150.
- Sida acuta Burm.f., 128, fig. 158.
    - cordifolia Linn., 128, fig. 154.
    - rhombifolia Linn., 128, fig. 157.
- Thespesia populnea (Linn.) Sol., 31, 39, 54, 191, fig. 38.
- Urena lobata Linn., 128, fig. 153.

BOTANICAL ARRANGEMENT OF THE SPECIES 271

Bombacaceae:
  Ceiba pentandra (Linn.) Gaertn., 158, 166.
  Durio zibethinus Murr., 154, fig. 210.
  Gossampinus malabaricus (DC.) Merr., 82.

Sterculiaceae:
  Commersonia bartramia (Linn.) Merr., 61, fig. 78.
  Heritiera littoralis Dry., 52, fig. 57.
  Helicteres hirsuta Lour., 61.
  Kleinhovia hospita Linn., 61, fig. 79.
  Melochia corchorifolia Linn., 128, fig. 156.
    umbellata (Houtt.) Stapf, 61.
  Pterospermum diversifolium Blume, 61.
  Sterculia foetida Linn., 29, 105, 186, fig. 34.
  Theobroma cacao Linn., 94, 166, fig. 209.
  Waltheria americana Linn., 128, fig. 152.

Dilleniaceae:
  Dillenia cauliflora Merr., 95.

Actinidiaceae:
  Saurauia angustifolia Becc., 100, fig. 97E.

Theaceae:
  Camellia sinensis (Linn.) O. Kuntze, 166.

Guttiferae:
  Calophyllum inophyllum Linn., 28, fig. 39.
  Garcinia linearis Pierre, 100, fig. 97C.
  Garcinia mangostana Linn., 153, fig. 211.

Bixaceae:
  Bixa orellana Linn., 159, 229, fig. 233.

Flacourtiaceae:
  Ahernia glandulosa Merr., 19.
  Flacourtia indica (Burm.f.) Merr., 155.
    inermis Roxb., 155.
    rukam Zoll. & Mor., 155.
  Pangium edule Reinw., 154, 187.

Passifloraceae:
  Passiflora edulis Sims, 152.
    foetida Linn., 129, fig. 160.
    quadrangularis Linn., 152.

Caricaceae:
  Carica papaya Linn., 152, 189, fig. 213.

Lythraceae:
  Pemphis acidula Forst., 32, 54, 191, fig. 36.

Sonneratiaceae:
  Sonneratia caseolaris (Linn.) Engl., 52, fig. 60.
Barringtoniaceae:
  Barringtonia asiatica (Linn.) Kurz, 29, fig. 41.
Rhizophoraceae:
  Bruguiera conjugata (Linn.) Merr., 51, fig. 53.
    cylindrica (Linn.) Blume, 51.
    parviflora (Roxb.) Wight & Arn., 51.
    sexangula (Lour.) Poir., 51.
  Ceriops tagal (Perr.) C. B. Rob., 51, fig. 54.
  Rhizophora apiculata Blume, 51, fig. 52A.
    mucronata Lam., 51, fig. 52B.
Combretaceae:
  Lumnitzera littorea (Jack) Voigt, 31, 54, fig. 43.
    racemosa Willd., 31.
  Terminalia catappa Linn., 29, 156, 186, fig. 42.
Myrtaceae:
  Eucalyptus deglupta Blume, 79.
  Eugenia mimica Merr., 100, fig. 97H.
  Melaleuca leucadendron Linn., 82.
  Psidium guajava Linn., 62, 152, 189, fig. 216.
  Syzygium aromaticum (Linn.) Merr. & Perr., 166.
    aqueum (Burm.f.) Alst., 154, fig. 217.
    cumini (Linn.) Skeels, 154, fig. 215.
    jambos (Linn.) Alst., 154, fig. 219.
    malaccense (Linn.) Merr. & Perr., 154.
    neriifolium Becc., 100, fig. 97D.
    samarangense (Blume) Merr. & Perr., 154.
Araliaceae:
  Polyscias nodosa (Blume) Seem., 110.
Myrsinaceae:
  Aegiceras corniculatum (Linn.) Blanco, 32, 54, fig. 61.
    floridum Roem. & Schult., 32, 54, fig. 62.
Plumbaginaceae:
  Plumbago auriculata Lam., 160, fig. 243.
    indica Linn., 160.
Sapotaceae:
  Achras zapota Linn., 152, 229, 230, fig. 218.
Oleaceae:
  Jasminum bifarium Wall., 34, 53, fig. 63.
    multiflorum Burm.f., 159.
    sambac (Linn.) Ait., 159, fig. 242.

Loganiaceae:
  Fagraea stenophylla Becc., 100, fig. 97B.
Apocynaceae:
  Allamanda cathartica Linn., 161, fig. 244.
  Catharanthus roseus (Linn.) G. Don, 37, 129, fig. 163.
  Cerbera manghas Linn., 32, fig. 40.
  Nerium indicum Mill., 160.
  Parameria laevigata (Juss.) Mold., 2.
  Plumeria acuminata Ait., 156, fig. 245.
  Thevetia peruviana (Pers.) K. Schum., 159, fig. 246.
Asclepiadaceae:
  Asclepias curassavica Linn., 128, fig. 162.
  Dischidia rafflesiana Wall., 98, fig. 93.
    vidalii Becc., 98, fig. 92.
  Hoya imbricata Decne., 98, fig. 91.
Convolvulaceae:
  Argyreia nervosa (Burm.f.) Boj., 161, fig. 247.
  Calonyction aculeatum (Linn.) House, 130.
    album (Linn.) House, 37.
  Erycibe longifolia Becc., 100, fig. 97F.
  Ipomoea aquatica Forsk., 228.
    cairica (Linn.) Sweet, 129.
    digitata Linn., 129.
    gracilis R. Br., 37.
    hederacea (Linn.) Jacq., 129.
    pes-caprae (Linn.) Roth, 37, 39, fig. 44.
    pes-tigridis Linn., 129.
  Merremia emarginata (Burm.f.) Hall.f., 129.
    gemella (Burm.f.) Hall.f., 129.
    hastata (Desr.) Hall.f., 129.
    hirta (Linn.) Merr., 129.
    umbellata (Linn.) Hall.f., 129.
    vitifolia (Burm.f.) Hall.f., 129.
  Quamoclit coccinea (Linn.) Moench, 129, fig. 164.
    pennata (Desr.) Boj., 129, fig. 165.
Boraginaceae:
  Cordia dichotoma Forst.f., 61.
    subcordata Lam., 32, fig. 45.
  Heliotropium indicum Linn., 130, fig. 161.
  Tournefortia argentea Linn.f., 32, fig. 47.
Verbenaceae:
  Avicennia marina (Forsk.) Vierh., 52, fig. 55.
  Clerodendron fistulosum Becc., 97, fig. 90.
    inerme (Linn.) Gaertn., 33.
    thomsonae Balf., 161, fig. 248.

Lantana camara Linn., 62, fig. 83.
Lippia nodiflora (Linn.) Rich., 130, fig. 166.
Petraea volubilis Linn., 161.
Stachytarpheta jamaicensis (Linn.) Vahl, 130, fig. 168.
Vitex negundo Linn., 228.
    trifolia Linn., 33.
    var. simplicifolia Cham., 32, 37, fig. 46.

Labiatae:
Anisomeles indica (Linn.) O. Kuntze, 130, fig. 167.
Hyptis capitata Jacq., 130.
    spicigera Lam., 130.
    suaveolens Poir., 130, fig. 171.
Leonurus sibiricus Linn., 130, fig. 170.
Leucas aspera (Willd.) Spreng., 130, fig. 181.
    lavandulifolia Sm., 130, fig. 178.
    zeylanica (Linn.) R. Br., 130, fig. 179.

Solanaceae:
Capsicum frutescens Linn., 131.
Cestrum nocturnum Linn., 105, 160.
Datura metel Linn., 131, fig. 177.
Lycopersicum esculentum Mill., 131.
Nicotiana tabacum Linn., 166.
Physalis minima Linn., 131, fig. 173.
    lanceifolia Nees, 131, fig. 172.
    peruviana Linn., 131, fig. 174.
Solanum nigrum Linn., 131, 192, fig. 169.

Scrophulariaceae:
Russelia juncea Zucc., 177, fig. 253.

Bignoniaceae:
Dolichandrone spathacea (Linn.f.) K. Schum., 52.
Spathodea campanulata Beauv., 156, fig. 250.
Tecoma stans (Linn.) Juss., 159, fig. 249.

Acanthaceae:
Acanthus ilicifolius Linn., 53, fig. 64.
Barleria cristata Linn., 159.
Graptophyllum pictum (Linn.) Griff., 158, fig. 252.
Thunbergia alata Boj., 131, fig. 175.
    erecta (Benth.) T. Anders., 160.
    grandiflora Roxb., 161, fig. 251.
    laurifolia Lindl., 161.

Rubiaceae:
Borreria laevis (Linn.) Griseb., 122.
Coffea arabica Linn., 166.
Dentella repens (Linn.) Forst., 131.
Guettarda speciosa Linn., 33, fig. 50.

Mitragyna speciosa Korth., 82.
Morinda citrifolia Linn., 33, 191, fig. 49.
Myrmecodia tuberosa Jack, 55, 96, fig. 96.
Neonauclea angustifolia (Havil.) Merr., 100, fig. 97A.
Psychotria acuminata Becc., 100, fig. 97J.
Scyphiphora hydrophyllacea Gaertn., 52

Cucurbitaceae:
  Benincasa hispida (Thunb.) Cogn., 149.
  Lagenaria siceraria (Mol.) Standl., 149.
  Luffa acutangula Roxb., 37, 149, fig. 212B.
    cylindrica (Linn.) M. Roem., 149, fig. 212A.
  Momordica charantia Linn., 148.

Goodeniaceae:
  Scaevola frutescens (Mill.) Krause, 33, fig. 48.

Compositae:
  Adenostemma lavenia (Linn.) O. Kuntze, 132, fig. 182.
  Ageratum conyzoides Linn., 132, fig. 176.
  Bidens pilosa Linn., 132, fig. 187.
  Chrysanthemum coronarium Linn., 228.
  Cosmos caudatus H.B.K., 132, fig. 188.
  Dichrocephala latifolia (Lam.) DC., 132, fig. 189.
  Eclipta alba (Linn.) Hassk., 132.
  Elephantopus scaber Linn., 132, fig. 186.
  Emilia sonchifolia (Linn.) DC., 192, fig. 184.
  Erechtites hieracifolia (Linn.) Raf., 132, fig. 185A.
    valerianaefolia (Wolf) DC., 132, fig. 185B.
  Erigeron sumatrensis Retz, 132.
  Pluchea indica (Linn.) Less., 37.
  Spilanthes acmella (Linn.) Murr., 192.
  Synedrella nodiflora (Linn.) Gaertn., 132, fig. 183.
  Vernonia cinerea (Linn.) Less., 132, fig. 180.
    patula (Ait.) Merr., 132.
  Wedelia biflora (Linn.) DC., 37, fig. 51.

Glossary

Acuminate; tapering to a point, the sides incurved, fig. 7B.
Acute; sharply pointed, the sides straight, fig. 7A.
Aggregate; a fruit composed of the many ripened carpels of a single flower.
Alternate; one after another, as leaves on different sides of a stem, one at each node.
Angiosperm; plants with their ovules in closed ovaries.
Annual; a plant that lives for a year or less and then dies.
Anther; that part of a stamen that produces the pollen.
Appressed; lying flat against the stem, or if a plant, then to the medium on which it grows.
Aril; a usually fleshy growth from the base of and more or less embracing the seed.
Axil; the angle on the upper side between a leafstalk and the stem.
Axillary; occurring in an axil.
Basal; appertaining to the base.
Berry; a fleshy, indehiscent, few- to many-seeded fruit.
Binomial; a combination of the Latin generic and specific names.
Bipinnate; twice pinnate.
Bryophyte; a moss plant.
Bulb; an underground stem composed of scales like the onion.
Calyx; the outermost envelope of the flower.
Capillary; hair-like in shape.
Capitulum; a dense mass of flowers, usually spherical in shape.
Capsule; a dry, one- to many-seeded seed-vessel developed from a compound ovary.
Carpel; a simple pistil or ovary.
Cauliflory; flowers and fruits borne on the trunk of a tree or shrub.
Cell; the cavity of an ovary or an anther.
Chlorophyll; the green coloring matter of leaves.
Collective; a fruit developed from the ovaries of several to many flowers.
Complete; having all parts.
Compound; similar parts aggregated into a whole.
Conical; having the form of a cone.

Cordate; heart-shaped, fig. 7F.

Coriaceous; resembling leather in texture.

Corm; a solid underground stem like that of the taro or the Jack-in-the-pulpit.

Corolla; the set of floral organs, next to the sepals, composed of petals and thus usually the showy part of a flower.

Cotyledon; the seed leaf.

Crenate; an edge with rounded teeth, fig. 8A.

Cryptogam; flowerless plants, such as the fungi, algæ, and ferns.

Deciduous; falling off; a term applied to leaves that fall at one time.

Dehiscent; splitting open.

Dentate; toothed margins, the teeth sharp and pointing outward, fig. 8B.

Depressed; flattened, as if pressed down from above.

Dicotyledon; plants with seeds that have two seed leaves.

Dimorphism; a term used to indicate the existence of two forms.

Drip tip; a slender tapering tip of a leaf to facilitate the removal of surface water.

Drupe; a fleshy fruit with a single hard seed such as a cherry or plum.

Ecology; the relations between organisms and their environment.

Elliptic; shaped like an ellipse, fig. 6F.

Endemic; confined to a restricted geographic region.

Endemism; the quality or state of being endemic.

Entire; the margins continuous, not toothed, lobed, or divided, fig. 8A.

Epiphyte; a plant growing upon another one but receiving no nourishment from its host.

Epiphytic; relating to epiphytes.

Evergreen; retaining leaves at all seasons.

Exotic; introduced from some other region.

Family; in botany, a natural group of plants having certain definite characters in common, such as the grasses, the palms, and the orchids.

Fascicle; a close cluster.

Female; a flower with a pistil but no stamens.

Filament; the stalk, that part of a stamen which supports the anther; the anther stalk.

Flaccid; limp, flabby.

Flagella; a long narrow and flexible growth from the end of some other organ.

Flora; all the plants of a particular region.
Frond; the leaf of a fern.
Generic; appertaining to a genus.
Genus; a group of plants having definite limiting characters in common, such as the maples (*Acer*), oaks (*Quercus*), pines (*Pinus*).
Glabrous; smooth, in the sense of having no hairs, bristles, or other covering.
Globose; spherical, like a ball.
Gregarious; growing together in mass.
Guttation; the dripping of internal moisture from a leaf tip.
Gymnosperm; a plant having naked ovules and naked seeds, the ovary wanting.
Halophytic; growing within the influence of salt water.
Head; a congested mass of small flowers, fig. 10E, F.
Heliophyte; plants adapted to growth in full sunlight.
Herb; a plant that is not woody.
Herbaceous; having the texture of an herb, as opposed to woody.
Hirsute; with stiff beard-like hairs.
Indehiscent; not splitting open.
Indigenous; native of a region as opposed to an introduced plant or animal.
Indumentum; any hairy or scaly covering.
Inequilateral; having unequal sides.
Inflorescence; the arrangement of the flowers on a plant.
Infructescence; the inflorescence in the fruiting stage.
Insular; appertaining to islands.
Internode; the part of a stem between two nodes.
Involucre; the bracts surrounding a flower or a group of flowers.
Irregular; asymmetric, fig. 9C.
Lanceolate; lance-shaped, fig. 6D.
Legume; a simple pod, one-celled, one- to many-seeded, as in the common bean.
Liana; a coarse woody vine.
Linear; narrow, many times as long as broad, the margins parallel, fig. 8C.
Lobe; any coarse projection or division of a leaf margin or other organ, fig. 8F.
Male; flowers with stamens but no pistil.
Membranaceous; thin and soft.
Monocotyledon; plants with a single seed leaf.

Monotypic; a term applied to a genus with a single species.

Mucro; a short abrupt tip.

Node; the joints of a stem; that part of stem where the leaves are attached.

Obovate; a flat inversely ovate body, the broad end upward; fig. 6B.

Obovoid; a solid body obovate in outline.

Obtuse; blunt, fig. 7C.

Opposite; leaves on opposite sides of the stem from each other at the same node.

Orbicular; circular in outline, fig. 6G.

Ovary; that part of the pistil containing the ovules, fig. 9A, E.

Ovate; like a longitudinal section of an egg, with the broader end downward, fig. 6A.

Ovoid; a solid body ovate in outline.

Ovule; the minute body within the ovary that is destined to become the seed, fig. 9A, H.

Palmate; arranged more or less like the outspread fingers of one's hand, fig. 8G, H.

Panicle; an open branched inflorescence, fig. 10C.

Pantropic; occurring in the tropics of both hemispheres.

Pappus; the tuft of hairs on the tips of certain seeds and small fruits.

Papyraceous; papery in texture.

Parasite; a plant growing upon another one and receiving some or all of its food from its host.

Peltate; a leaf with its stalk attached on the lower surface within the margin; shield-shaped, fig. 7G.

Perennial; lasting from several to many years.

Perfect; a flower with both stamens and pistils, fig. 9A, B.

Pericarp; the fruit walls.

Persistent; remaining beyond the period when such parts usually fall.

Petal; a single part of the corolla, fig. 9A, B.

Petiole; leafstalk.

Phanerogam; a plant bearing flowers and producing seeds.

Phytogeographic; appertaining to plant geography; the distribution of plants.

Pinnate; a compound leaf where the leaflets are arranged along both sides of a common stalk, like a feather, fig. 8I.

Pistil; the female organ of a flower consisting of the ovary, ovules, style, and stigma, fig. 9A, E, F.

GLOSSARY

Pistillate; having a pistil.
Pod; a term applied especially to the fruit of the bean family.
Pollen; the normally yellow powder produced within the anther that fertilizes the ovules.
Pollinate; the act of transferring pollen from the anther to the stigma.
Polytypic; a genus with few to many species.
Primary forest; the virgin forest unaltered by the activities of man.
Prostrate; lying flat on the ground.
Pubescent; hairy.
Pyriform; pear-shaped.
Raceme; an unbranched, elongated inflorescence, the flowers all stalked, fig. 10B.
Rachis; the axis of an inflorescence.
Radicle; the primary root.
Reflexed; bent backward.
Regular; all parts of each set of organs similar in shape and size, fig. 9B.
Retrorse; abruptly reflexed.
Retuse; an obtuse tip with a small indentation, fig. 7E.
Rhizome; an underground stem resembling a root but bearing buds.
Rootstock; a rhizome.
Ruderal; growing in waste places.
Saprophyte; a plant without chlorophyll, growing on decayed vegetable matter.
Scandent; climbing.
Secondary forest; that type composed of fast-growing small trees that usually develops when the primary forest is destroyed.
Semi-; in compound words of Latin origin, half, as *semiparasitic*, half parasitic.
Sepal; a division of the calyx, fig. 9A.
Serrate; a margin cut into sharp teeth, the teeth pointing upward, fig. 8C.
Sessile; without any stalk.
Shrub; a woody plant, with or without a trunk, but 12 to 15 feet high or less.
Simple; single, as opposed to compound.
Solitary; single.
Spadix; a fleshy spike bearing numerous small flowers, as in the central part of the calla lily.
Spathe; the organ that encloses an inflorescence, as in the calla lily.

Species; a group of plants having definite limiting characters within a larger group (genus).
Specific; relating to species.
Spherical; globose, like a ball.
Spike; an inflorescence like a raceme but the flowers all sessile, fig. 10A.
Spore; the microscopic reproductive cells of the flowerless plants, fungi, mosses, and ferns.
Stamen; the male organs of the flower, fig. 9A, C.
Staminate; furnished with stamens.
Stenophylly; a term indicating the character of having narrow leaves.
Sterile; barren or imperfect.
Stigma; that part of the pistil that receives the pollen, fig. 9A, G.
Strand; appertaining to the seashore, especially open beaches.
Style; the stalk between the ovary and the stigma, fig. 9A, F.
Succulent; fleshy.
Symbiosis; the living together of dissimilar organisms.
Terete; cylindric.
Terrestrial; appertaining to the earth.
Tomentose; clothed with matted woolly hairs.
Transpiration; the evaporation of water from the plant tissues and its passage into the air.
Tree; a woody plant with a definite trunk 15 to 20 feet high or more.
Tripinnate; thrice pinnate.
Tuber; a thickened underground stem with lateral buds, like the potato.
Tubercle; a small excrescence.
Umbel; the umbrella form of inflorescence, fig. 10D.
Undulate; wavy-margined, fig. 8E.
Variety; a category under the species to indicate minor variations.
Venation; the arrangement of veins in leaves.
Verticillate; arranged in a whorl.
Viability; the ability of seeds to retain life over various periods of time.
Villose; clothed with long soft hairs.
Vine; any plant with a trailing or twining stem.
Whorl; an arrangement of leaves, or other organs, in circles around the stem on axis.

Index

A

Aátasi, 36
Abacá, 166
Abelmoschus moschatus, 128, 141, 151, 228
Abrus precatorius, 127, 136
Abutilon indicum, 128, 140
Acacia farnesiana, 62, 69
Acalypha hispida, 158, 174
 indica, 128, 139
 wilkesiana, 158, 174
Acanthus ilicifolius, 53, 58
Achióte, 159
Achras zapota, 152, 171, 229, 230
Achuéte, 229
Achyranthes aspera, 126
Acrostichum aureum, 52, 58, 108, 181
Adenanthera pavonina, 157
Adenostemma lavenia, 132, 145
Aegiceras corniculatum, 32, 54, 58
 floridum, 32, 54, 58
African oil palm, 162
 tulip-tree, 156
Ágel, 163
Ageratum conyzoides, 132, 144
Aglaia, 79
Agóho, 28
Agútag, 32
Ahernia glandulosa, 19
Aímoa, 36
Aírie, 36
Akánkan, 36
Akan-paku, 108
Akapúlko, 230
Alagáo, 61
Álang-álang, 65
Albizzia lebbek, 157
Aleurites moluccana, 61, 70
Aleutian Islands flora, 232, 250
Algae, 10
Alím, 61
Alipáta, 31
Allamanda cathartica, 161, 176

Allium tuberosum, 228
Allophylus, 33
Almond, Indian, 156, 186
Alocasia macrorrhiza, 104
Alternanthera sessilis, 126, 134
 versicolor, 165
Amaranthus spinosus, 126, 135
 tricolor, 165
 viridis, 126, 135
Amargóso, 149
Amorphophallus campanulatus, 103
 titanum, 103
Ampaláya, 149
Anacardium occidentale, 151, 169, 189
Ananas comosus, 151
Andropogon sorghum, 147
Angiopteris, 106
 evecta, 110
Angiospermae, 10, 11
Angsána, 157
Aníbong, 163, 183
Anílao, 61
Anisomeles indica, 130, 143
Anisoptera, 80
Annatto, 159, 229
Annona muricata, 152, 167, 230
 reticulata, 152, 167
 squamosa, 152, 167
Anónang, 61
Anónas, 152
Antiaris toxicaria, 1
Antidesma, 188
 ghaesembilla, 61, 70
Antigonon leptopus, 161, 172
Antrophyum plantagineum, 107, 110
Ápi-ápi, 52
Apíri, 31
Apium graveolens, 228
Aposótis, 229
Araucaria, 84
Areca catechu, 162, 164
Áren, 164, 183
Arenga pinnata, 164, 183

283

INDEX

Argemone mexicana, 126, 135
Argyreia nervosa, 161, 176
Aróma, 62
Artocarpus altilis, 167, 188
    champeden, 95, 155
    heterophylla, 95, 153, 228
    odoratissima, 155
    rotunda, 155
Asclepias curassavica, 128, 142
Asplenium nidus, 55, 106
Atalantia linearis, 100
Áte-áte, 37
Athyrium esculentum, 181
Átis, 152
Atóto, 36
Australian flora, 256
Averrhoa bilimbi, 153, 169
    carambola, 94, 153, 169
Avicennia marina, 52, 57

B

Babakóan, 32
Baccaurea racemosa, 155
Báchang, 151, 189
Badíang, 104
Bádu, 103
Bagnít, 53
Bágo, 62, 187
Bakáu, 51
Bakaúan, 51
Bákong, 35
Balanophora, 101
Balánti, 61
Balábas, 158
Balatbás, 35
Balíknon, 61
Bálok-bálok, 32
Balongái, 61
Bamboos, 161
Banálo, 31, 191
Banana, 150, 190
Banáto, 61
Bangkóng, 32
Bangkúdo, 191
Bankúdo, 33
Bantígi, 32
Banyan, 158
Barleria cristata, 159
Barringtonia asiatica, 29, 46

Batai, 157
Bátao, 228
Batáoan, 148
Batikúling, 61
Bau, 30
Bauhinia acuminata, 160, 173
    monandra, 160
    purpurea, 160
    tomentosa, 160, 173
Bayóg, 61
Beech, 15
Benincasa hispida, 149
Bermuda-grass, 123
Bétah, 31
Betel nut palm, 162
    pepper, 164
Bidáru, 31
Bidens pilosa, 132, 146
Binyayóyo, 61
Bínjai, 151, 188
Binomials, 16
Binónga, 61
Bintáru, 32
Bird's nest fern, 55, 106
Bixa orellana, 159, 174, 229
Blának, 32
Blímbing, 153
Bóboy, 157
Boerhavia diffusa, 35, 41, 179, 192
Borassus flabellifer, 163
Borneo flora, 256
Borreria laevis, 122
Botanical classification, 9
Bóto, 33
Bótong, 29
Bottle gourd, 149
Bougainvillea spectabilis, 161, 172
Bowstring-hemp, 159
Brassica pekinensis, 228
Breadfruit, 188
Brown algae, 10
Bruguiera conjugata, 51
    cylindrica, 51
    parviflora, 51
    sexangula, 51
Bryophyta, 10
Búa-búa, 33
Búah-nóna, 152
Bullock's-heart, 152
Bunga, 162

INDEX

Búri, 91, 163, 183
Burmannia, 102
Busáin, 51
Búta-búta, 31
Buttercup, 87

C

Cabellero, 159
Cacao, 94, 166, 229
Cactaceae, 13
Cadena de amor, 161
Caesalpinia crista, 53
　nuga, 53, 57
　pulcherrima, 159, 173
Cajanus cajan, 148, 168
Caladium bicolor, 165
Calamus, 73
Calonyction aculeatum, 130
　album, 37
Calophyllum, 79
　inophyllum, 28, 46
Camellia sinensis, 165
Campnosperma macrophylla, 82
Cana fistula, 157
Cananga odorata, 156, 173
Canarium, 62
　commune, 68, 187
　ovatum, 187
Canavalia maritima, 36, 43
　microcarpa, 37, 43
Canna indica, 125
Capsicum frutescens, 131
Carambóla, 153
Cardiospermum halicacabum, 127, 139
Carica papaya, 152, 171, 189
Cashew, 151, 189
Cassava, 149
Cassia alata, 126, 136, 230
　fistula, 157, 174
　javanica, 158
　nodosa, 157
　occidentalis, 126, 136
　siamea, 157, 174
　tora, 126, 127, 137
Cassytha filiformis, 36, 42
Castanea, 15
Castanopsis, 80, 84
Casuarina equisetifolia, 28, 29, 41
Catharanthus roseus, 37, 128, 142

Cauliflory, 94
Ceiba pentandra, 158
Celosia argentea, 126, 134
　cristata, 165
Cerbera manghas, 32, 46
Cereals, 147
Ceriops tagal, 51
Cestrum nocturnum, 105, 160
Champáka, 157, 228
Cheiropleuria bicuspis, 109, 110
Chempáka, 157
Chempédak, 95, 155
Chenopodium ambrosioides, 126, 136, 229
Chestnut, 15
　Polynesian, 155, 186
Chíko, 152, 229
Chile pepper, 131
Chinese flora, 251
Chlorophyceae, 10
Chocolate, 94, 229
Chrysanthemum coronarium, 228
Cinnamomum, 79
Clerodendron inerme, 33
　fistulosum, 97
　thomsonae, 161, 177
Clitorea ternatea, 138
Clove, 166
Cockscomb, 165
Coconut, 34
　palm, 162
Cocos nucifera, 34
Codiaeum variegatum, 159, 175
Coffea, 165
　arabica, 165
Coix lachryma-jobi, 147
Colocasia esculenta, 149, 167
Colona, 61
Colubrina asiatica, 33, 45
Commelina benghalensis, 125, 134
　diffusa, 125, 134
Commersonia bartramia, 61, 69
Compositae, 14
Conchophyllum, 98
Coniferae, 11
Corchorus acutangulus, 128
　capsularis, 128
　olitorius, 128, 140
Cordia dichotoma, 61
　subcordata, 32, 47

Cordyline fruticosa, 159, 172
Corypha, 91
  elata, 163
Cosmos caudatus, 132, 146
Couthovia, 212
Cowpea, 148
Crinum asiaticum, 35, 40
Crotalaria incana, 127, 138
  mucronata, 127, 137
  quinquefolia, 127, 138
  retusa, 127, 138
  verrucosa, 127
Croton, 159
  tiglium, 160
Cryptocarya, 79
Cubebs, 166
Cucurbitaceae, 14
Cultivated plants, 147
Custard-apple, 152
Cyanotis cristata, 125, 134
Cyathea, 106
Cycadaceae, 11
Cycas circinalis, 30, 40, 187
Cyclophorus lanceolatus, 107, 110
Cymbidium, 111
Cynodon dactylon, 123
Cynometra cauliflora, 94, 155
Cyperaceae, 11
Cypress-vine, 129
Cypripedium, 111

Dichrocephala latifolia, 132, 146
Dicotyledoneae, 11
Dilíman, 52, 108, 181
Diliúario, 53
Dillenia cauliflora, 95
Dinochloa, 86
Dioscorea alata, 149, 167, 185, 224
  esculenta, 61, 68, 149, 185
  hispida, 61, 68
Diospyros, 79
Dipteris conjugata, 109, 110
Dipterocarpus, 80
Dischidia rafflesiana, 98
  vidalii, 98
Dishcloth gourd, 149
Dodonaea viscosa, 31, 44
Dolichandrone spathacea, 52
Dolicholobium, 212
Dolichos lablab, 148, 168, 228
Doloáriu, 53
Drosera, 3
Drymoglossum carnosum, 109, 110
Drynaria quercifolia, 54, 108, 109
Dúgtong áhas, 2
Dúhat, 154
Dúku, 94, 154
Dúngon-láte, 52
Durián, 154
Durio zibethinus, 154, 170
Dysoxylum, 79

D

Daemonorops, 73
Dágo, 185
Dama de noche, 105, 160
Dampálit, 36
Dap-dap, 29, 191
Dápo, 115
Dápo-maripósa, 111
Datura metel, 131, 144
Dáua, 228
Delonix regia, 156, 174
Dendrobium, 111
Dentella repens, 131
Derris trifoliata, 53, 57
Desmodium gangeticum, 137
  pulchellum, 137
  umbellatum, 32, 42, 53
  velutinum, 137

E

Eclipta alba, 123, 132
Elaeis guineensis, 162
Elaeocarpus, 79, 216
Elephantopus scaber, 132, 146
Eleusine corocana, 147
  indica, 123
Emilia sonchifolia, 132, 146, 192
Endospermum moluccanum, 97
Entada phaseoloides, 34, 103
Epirixanthes, 102
Érang, 164
Erechtites hieracifolia, 132, 146
  valerianaefolia, 132, 146
Eria, 111
Erigeron canadense, 123
  sumatranum, 132
Erycibe longifolia, 100

INDEX

Erythrina variegata, 29, 43, 191
Eucalyptus deglupta, 80
Eugenia, 79
  mimica, 100
Euphorbia atoto, 36, 44
  heterophylla, 127, 139
  hirta, 127, 139
  pulcherrima, 160, 175
  prostrata, 127
  thymifolia, 127
Excoecaria agallocha, 5, 31, 44, 54

F

Fádang, 30, 187
Fagraea stenophylla, 100
Fagus, 15
Fau, 30
Ferns, 10, 106
  edible, 181
Ficus, 79
  benjamina, 158
  elastica, 158
  mirabilis, 96
  religiosa, 158
Figs, strangling, 93
Fiji flora, 253
Fire-tree, 156
Fish-tail palm, 164
Flacourtia inermis, 155
  rukam, 155
Flagellaria indica, 35
Flamboyant, 156
Flame-tree, 156
Flemingia strobilifera, 137
Flora of:
  Aleutians, 232, 250
  Australia, 251
  China, 251
  Christmas, 253
  Fanning, 253
  Fiji, 253
  Formosa, 235, 252
  Funafuti, 253
  Galápagos, 239, 252
  Guam, 253
  Hainan, 235, 252
  Hawaii, 239, 252
  Indo-China, 252
  Japan, 235, 251
  Java, 256
  Johnston, 253
  Kamchatka, 251
  Kuriles, 233, 251
  Makatea, 254
  Malay Archipelago, 235, 256
  Malay Peninsula, 256
  Marquesas, 253
  Micronesia, 253
  New Caledonia, 255
  New Guinea, 255
  New Zealand, 256
  Niue, 253
  Pacific Islands, 240, 252, 253
  Palmyra, 253
  Philippines, 235, 256
  Rarotonga, 253
  Riu Kiu, 253
  Sakhalin, 251
  Samoa, 253
  Tahiti, 253
  Tonga, 253
  Wake, 253
Flowering plants, 10
Flowers, characters of, 22
  types of, 23
Forests, deciduous, 82
  dipterocarp, 81
  endemism in, 81
  evergreen, 82
  mangrove, 49
  mid-mountain, 83
  mossy, 85
  primary, 71
  secondary, 60
Formosan flora, 235
Frangipani, 156, 229
Freycinetia, 73
Fruits, characters of, 24
  edible, 86
  tropical, 150
  types of, 25
Fúe, 36
Funafuti flora, 253

G

Gábi, 149
Galápagos Islands flora, 239, 252
Galeola, 102, 112
Gandasúli, 228

Gápas, 228
Garcinia, 79
    linearis, 100
    mangostana, 153, 170
Gébang, 163, 183
Gelágah, 65
Gélang, 36
Gélang-laut, 36
Gendíwung, 164
Gendúru, 163, 183
Generic names, 16
Gentian, 87
Gleichenia, 86, 110
Gliricidia sepium, 229
Gluta, 4
Gnetum gnemon, 62, 68, 187
Gógo, 34, 103
Golasíman, 36
Golden-shower, 157
Gomaméla, 159
Goniothalamus, 79
Gossampinus malabaricus, 83
Gramineae, 11
Grammatophyllum speciosum, 113
Granadilla, 152
Graptophyllum pictum, 158, 177
Grasses, edible, 184
Green algae, 10
Green gram bean, 148
Guam flora, 253
Guanábanos, 152
Guava, 62, 152, 189
Guettarda speciosa, 33, 48
Gunpowder plant, 126
Gymnospermae, 10
Gynandropsis gynandra, 126, 135
Gyrocarpus americanus, 82

H

Hagnáya, 52, 108
Hagónoi, 37
Hainan flora, 235, 252
Hanagdóng, 60
Handámo, 33
Handjúwang, 159
Hau, 30
Hawaiian flora, 239, 252
Hedychium coronarium, 228
Helicteres hirsuta, 61
Heliotropium indicum, 130, 142

Helminthostachys zeylanica, 108, 109
Hepaticae, 10
Heritiera littoralis, 52, 57
Hernandia ovigera, 29, 41
    peltata, 29
Hibiscus mutabilis, 159, 175
    rosa-sinensis, 159, 175
    schizopetalus, 159, 175
    surattensis, 128, 141
    tiliaceus, 17, 30, 45, 54
Hidiók, 164
Hína, 190
Hinlaúmo, 61
Homalanthus populneus, 61, 69
Homonoia riparia, 100
Hopea, 80
Horse-radish-tree, 158
Horsfieldia, 79
Hoya imbricata, 98
Hyacinth bean, 148, 228
Hymenolepis spicata, 107
Hyptis brevipes, 130
    capitata, 130
    spicigera, 130
    suaveolens, 130, 143

I

Ídiok, 183
Ífi, 186
Íhi, 186
Ilangilang, 156
Imperata cylindrica, 65
India-rubber tree, 158
Indian almond, 156, 186
Indian corn, 147
Indigofera hirsuta, 136
    suffruticosa, 136
    tinctoria, 136
Indo-China flora, 252
Inflorescences, types of, 24
Inocarpus fagiferus, 155, 169, 186
Insectivorous plants, 3
Ipél, 62
Ipomoea aquatica, 228
Ipomoea cairica, 129
    digitata, 129
    gracilis, 37
    hederacea, 129
    pes-caprae, 37, 47
    pes-tigridis, 129

Ischaemum muticum, 34
Ixora, 159

J

Jakfruit, 95, 153
Jambé, 162
Jambolán, 154
Jámbu áyer, 154
Jámbu bol, 154
Jámbu semárang, 154
Japanese flora, 233, 251
Jasmine, 159
Jasminum bifarium, 34, 53, 58
  multiflorum, 159
  sambac, 159, 176
Jatropha curcas, 61, 70, 160
Jimson-weed, 131
Job's tears, 147
Jungermanniaceae, 10

K

Kabatíti, 33
Kábong, 164
Kachúmba, 228
Kádios, 148
Kakuáte, 229
Kalachúchi, 156, 229
Kalamísmis, 148
Kalapíni, 37
Kalumbibít, 53
Kalúmpang, 29, 105, 186
Kamamchíli, 229
Kamáte, 229
Kamchatkan flora, 251
Kamóte, 150, 229
Kamóting-káhoy, 149
Kamúning, 159
Kanári, 187
Kandelia, 51
Kankóng, 228
Káong, 164, 183
Kápas, 228
Kapok, 157, 166
Kapulásan, 153
Kapúrko, 230
Karambódja, 228
Karóti, 61
Kastúli, 228
Kátjang, 148

Kátjang manila, 230
Katúra, 190
Katúrai, 191
Katúray, 157
Katúri, 35
Káyu-járau, 52
Kébing-bangáh, 103
Kembóla, 94
Kenánga, 156
Kepáyang, 154
Keráskas, 52
Ketápang, 156, 186
Ketchápi, 154
Ketjeper, 148
Kleinhovia hospita, 61, 70
Knema, 79
Kógon, 65
Kolítis, 229
Kólung-kólung, 29
Kondól, 149
Korthalsia, 97
Krúbi, 103
Kubíli, 36
Kúchai, 228
Kúnai, 65
Kurile Islands flora, 233, 251
Kwíni, 151, 189

L

Labáya, 61
Labiatae, 14
Láfo, 35
Lagenaria siceraria, 149
Lagólo, 52, 108, 181
Lagúndi, 32, 33, 228
Lálang, 65
Lamídiang, 181
Lángsat, 94, 154
Lánguil, 157
Lánjut, 151, 188
Lansium domesticum, 94, 154, 169
Lansóne, 94, 154
Lantana camara, 62, 70
Laportea, 5
Lasóna, 228
Leaves, characters of, 20
  types of, 20
Lecanopteris, 86
Lecanopteris carnosa, 99, 110
Leguminosae, 14

Lei-lei, 31
Lembíding, 52
Leonurus sibiricus, 130, 143
Leucaena glauca, 62, 69
Leucas aspera, 130, 145
  lavandulifolia, 130, 145
  zeylanica, 130, 145
Líbas, 151
Licuala spinosa, 35
Lígas, 4
Liliaceae, 11
Lima bean, 148
Lipáta, 32
Lippia nodiflora, 130, 143
Litchi chinensis, 153
Lithocarpus, 80, 84
Litsea, 61, 79
Liverworts, 10
Líwung, 164
Loofah, 149
Lovi-lovi, 155
Luffa acutangula, 149, 170
  cylindrica, 37, 149, 170
Lumanái, 100
Lúmbang, 61
Lumbía, 164
Lumnitzera coccinea, 54
  littorea, 31, 46
  racemosa, 31
Lychee, 153
Lycopersicum esculentum, 131
Lygodium circinnatum, 107

M

Macaranga caladifolia, 97
  tanarius, 61, 68
Madre de cacao, 229
Maize, 147
Makatea flora, 253
Makópa, 154
Malabágo, 30
Malapígas, 32
Malasápsap, 82
Malay Peninsula flora, 256
Malayan floras, 235, 256
Malay-apple, 154
Malaysian plant distribution, 193
Malisa, 228
Mallotus philippensis, 61, 69
  ricinoides, 61, 69

Malúngai, 158, 191
Malvastrum coromandelinum, 128, 140
Mangifera, 4
  caesia, 151, 188
  foetida, 151, 189
  indica, 151, 169
  lagenifera, 151, 188
  odorata, 151, 189
Mango, 4, 151
Mangosteen, 153
Mangrove forests, 49
Manihot esculenta, 169
Manila hemp, 166
Manól, 34
Mapóla, 159
Márang, 155
Marquesas flora, 253
Masóa, 35
Melaleuca leucadendron, 82
Melanolepis multiglandulosa, 61, 70
Melanorrhoea, 4
Melochia corchorifolia, 128, 141
  umbellata, 61
Melochyla, 4
Mempári, 32
Menteng, 155
Mentígi, 32
Merremia emarginata, 129
  gemella, 129
  hirta, 129
  umbellata, 129
  vitifolia, 129
Metroxylon, 82
  sagu, 164
Mexican poppy, 126
Michelia alba, 157
  champaca, 157, 173, 228
Micronesian floras, 253
Milkweed, 128
Millet, 147
Mílo, 31
Mimosa pudica, 127, 138
Míro, 31
Mitragyne speciosa, 82
Moláve, 61
Molitái, 31
Momordica charantia, 149
Mondóa, 190
Monocotyledoneae, 11

INDEX

Moonflower, 37
Morinda citrifolia, 33, 48, 191
Moringa oleifera, 158, 191
Mosses, 10
Motíni, 37
Mucuna albertisii, 102
Muehlenbeckia platyclada, 160, 172
Múngos, 148
Murmasáda, 32
Murraya paniculata, 159
Musa textilis, 166
    uranoscopos, 110
Myrmecodia tuberosa, 55, 96
Myrmeconauclea, 97
Myristica, 79
    fragrans, 166
    myrmecophila, 97

N

Náli, 187
Námi, 61
Nam-nam, 94, 155
Nanéa, 36
Nángka, 95, 153
Nángka-blánda, 152
Nárra, 157
Neonauclea, 97
    angustifolia, 100
Nepenthes, 3, 86
Nephelium lappaceum, 153, 170
    mutabile, 153, 170
Nerium indicum, 160
Nettle, 5
New Caledonia flora, 255
New Guinea cabbage, 62
    flora, 255
New Hebrides flora, 255
New Zealand flora, 256
Nicotiana tabacum, 166
Nílad, 52
Nílar, 52
Níno, 33
Nipa, 34
    fruticans, 54, 164
    palm, 164
Niue flora, 253
Nothopanax, 158
Nutmeg, 166

O

Oak, 15
Oleander, 160
Oleandra, 110
Oncosperma filamentosum, 164
Ophioglossum pendulum, 107, 108
Opo, 149
Orchidaceae, 11
Orchids, 111
Oryza sativa, 147
Oúru, 33
Oxalis corniculata, 126
    repens, 16, 138

P

Pachyrrhizus erosus, 148, 168
Pacific Islands floras, 240, 252, 253
Pagátpat, 52
Pákis-bang, 52
Pakis-kalér, 108
Páko, 181
Pakó-laut, 108
Pákpak-laúin, 54, 108
Páku-laut, 52, 181
Palaquium, 79
Palmae, 11
Palms, 162
    edible buds of, 182
    starch from, 183
Palmyra palm, 163
Pálo maria, 28
Pána, 228
Pándan, 30, 54, 179, 189
Pandanus brosimos, 190
    conoideus, 190
    julianetti, 190
    tectorius, 29, 40, 54, 179, 190
Pangium, 37
    edule, 154, 187
Pángui, 154, 187
Papúa, 158
Papáya, 152, 189
Parameria laevigata, 2
Parasites, 101
Parkia roxburghii, 18
Passiflora edulis, 152
    foetida, 128, 142
    quadrangularis, 152

INDEX

Passion-fruit, 152
Patáning-dágat, 36
Patóla, 37, 149
Péchai, 228
Pelótok, 32
Peltophorum pterocarpum, 157
Pemphis acidula, 32, 45, 54
Pepper, 166
  Chile, 131
Petraea volubilis, 161
Péye, 52
Phaeophyceae, 10
Phalaenopsis amabilis, 111
Phanerogamia, 10
Phaseolus aureus, 148
  calcaratus, 148
  lunatus, 148, 168
Philippine flora, 235, 256
Photinopteris speciosa, 109, 110
Phyllanthus niruri, 128, 139
  urinaria, 128
Physalis lanceifolia, 131, 144
  minima, 131, 144
  peruviana, 131, 144
Physic-nut, 61, 160
Pía, 35
Piágau, 31, 52
Pigeon pea, 148
Pilápil, 32, 54
Pilea microphylla, 126
Píli, 187
Pínang, 162
Pineapple, 151
Pink-shower, 157
Pípal, 158
Piper betle, 164
  cubeba, 166
  myrmecophilum, 97
  nigrum, 166
Pirítai, 36
Pítai, 36
Pitcher plant, 3
Pithecellobium dulce, 229
Pitógo, 30
Plant families, 13
Plants, cultivated, 147
  edible, 147, 178
Platycerium, 108
Pluchea indica, 37
Plumbago auriculata, 160, 176

Plumeria acuminata, 156, 176
  acutifolia, 229
Pohúehúe, 37
Poinsettia, 160
Poison ivy, 4
Poisonous plants, 4
Polanisia icosandra, 126, 135
Polianthes tuberosa, 165
Polyalthia, 79
Polynesian chestnut, 155, 186
  plant distribution, 205
Polypodium heracleum, 108
  sinuosum, 55, 99, 110
Polyscias nodosa, 110
Pomelo, 151
Pongamia pinnata, 32, 42
Portulaca oleracea, 36, 126, 135, 179, 191
Póso, 150, 190
Potentilla, 87
Poúe, 36
Pouzolzia zeylanica, 126
Premna, 61
Pseuderanthemum, 158
Psidium guajava, 62, 152, 171, 189
Psophocarpus tetragonolobus, 148, 168
Psychotria, 61
  acuminata, 100
Pterocarpus indica, 157
Pteridophyta, 10
Pterospermum diversifolium, 61
Pugáhan, 163, 183
Pulásan, 153
Pungápung, 103
Purslane, 179, 191
  seaside, 179, 191

Q

Quamoclit coccinea, 129, 142
  pennata, 129, 142
Quercus, 15, 80
Quinine, 165

R

Ragi, 147
Rain-tree, 156
Rámbai, 155

INDEX

Rafflesia arnoldi, 101
  schadenbergiana, 101
Rambútan, 153
Rarotonga flora, 253
Raspberry, 86
Rattan palm, 73, 162
Ravenala madagascariensis, 156, 172
Renanthera, 111
Red algae, 10
Remirea maritima, 34
Réngas, 4
Rhizophora apiculata, 51
  mucronata, 16, 51
Rhododendron, 86
Rhodophyceae, 10
Rhoeo discolor, 165, 172
Rhus, 4
Rice, 147
  bean, 148
Ricinus communis, 139
Ríma, 35
Riu Kiu Islands flora, 235
Rosaceae, 13, 15
Rose-apple, 154
Royal palm, 164
Rúkam, 155
Rumbía, 164
Rúmpet kerúpet, 34
Russelia juncea, 177

S

Saccharum officinarum, 148
  spontaneum, 65
Sádang, 163
Sága, 127, 157
Sago palm, 82, 164
Ságu, 164
Saguilála, 159
Sakhalin flora, 251
Salakáho, 32
Samanea saman, 156
Saminánga, 33
Samoa flora, 253
Sampaguíta, 159
Sampálok, 157
Sandoricum koetjape, 154
Sansevieria roxburghiana, 159
Santól, 154
Sapínit, 53

Sapodilla, 229, 230
Saprophytes, 102
Sárai, 163
Sarása, 158
Saurauia, 98
  angustifolia, 100
Sáwo maníla, 152, 230
Scaevola frutescens, 33, 48
Schizophyta, 10
Schizostachyum, 86
Screw-pine, 30, 189
Scyphiphora hydrophyllacea, 52
Seashore plants, 27
Seaside plants, 27
  purslane, 179, 191
Seed plants, 10
Seeds, characters of, 26
  edible, 186
  types of, 26
Semecarpus, 4
  gigantifolius, 95
Sensitive plant, 127
Sentígi, 32
Sérdang, 163
Sesbania grandiflora, 173
  sesban, 157, 191
Sesuvium portulacastrum, 36, 42, 179, 191
Setaria italica, 147, 228
Shaddock, 151
Shorea, 80
Shrubs, ornamental, 159
Siántan, 159
Sida acuta, 128, 141
  cordifolia, 128, 141
  rhombifolia, 128, 141
Sínkamas, 148
Sirikája, 152
Sisal, 166
Sítao, 148, 228
Sloanea, 79
Solanum nigrum, 131, 143, 192
Sonneratia caseolaris, 52, 57
Sophora tomentosa, 32, 43
Sorghum, 147
Soursop, 152, 230
Spathodea campanulata, 156, 177
Spermatophyta, 10
Spinach, substitutes for, 191
Spinifex littoreus, 34, 37, 40

Spondias cytherea, 151
  dulcis, 188
  pinnata, 151, 188
Stachytarpheta jamaicensis, 130, 143
Stag-horn fern, 108
Stenochlaena palustris, 53, 58, 108, 182
Sterculia foetida, 29, 44, 105, 186
Stinging plants, 5
Strand plants, 27
Strongylodon, 102
Strychnos, 7
Sugar cane, 148
Sundew, 3
Suriana maritima, 33, 42
Suwángkung, 163
Sweet potato, 150, 229
Sweetsop, 152
Swintonia, 4
Synedrella nodiflora, 132, 145
Syzygium, 79
  aqueum, 154, 171
  aromaticum, 166
  cumini, 154, 171
  jambos, 171
  malaccense, 154
  neriifolium, 100
  samarangense, 154

T

Tábau, 31, 54
Tabígi, 52
Tabóbok, 37, 149
Tabúg, 33
Tacca pinnatifida, 35, 41, 185
Taeniophyllum, 113
Tagalínau, 83
Tagárai, 36
Takpó, 61
Taláhib, 65
Tálie, 149
Talipot, 91
Táli-putéri, 36
Talísay, 29, 156, 186
Talósan, 61
Tamarind, 157
Tamarindus indica, 157, 168
Tambalísa, 32
Támis, 149
Tanág, 61

Tangál, 51
Tángkal-ásem, 157
Tapioca, 149
Taro, 149
Tea, 165
Tecoma stans, 159, 177
Temple-tree, 156
Tempúni, 155
Terminalia, 79
  catappa, 29, 46, 156, 186
Terúlak, 37
Teruntum, 31, 54
Tetranthera salicifolia, 100
Thallophyta, 9
Theobroma cacao, 94, 170
Thespesia populnea, 31, 45, 54
Thevetia peruviana, 159, 176
Thuarea involuta, 34
Thunbergia alata, 144
  erecta, 160
  grandiflora, 161, 177
  laurifolia, 161
Tíkas-tíkas, 125
Tikúsan, 34
Titíti, 36
Tobacco, 166
Tomato, 131, 229
Tonga flora, 253
Tónkud-lánguit, 108
Tóto, 36
Tournefortia argentea, 32, 47
Traveler's-tree, 156
Tree nettle, 5
Trees, ornamental, 156
Trema orientalis, 60, 68
Tristellateia australasiae, 53, 58
Triumfetta bartramia, 128, 140
  procumbens, 36, 44
  repens, 36
  semitriloba, 128
Túba, 160
Túba-laut, 53
Tubers, edible, 185
Túe, 52
Túgi, 61
Túgue, 149
Tulip-tree, 156
Tungháo, 228
Túngo, 61, 149, 185
Turúka, 35

U

Uag, 35
Ubi, 149, 185
Ugágoi, 33
Ugsang, 35
Ungsoi, 228
Upas tree, 1
Urena lobata, 128, 140
Urtica, 5

V

Vaccinium, 86
Vanda, 111
Vatica, 80
Vau, 30
Vavaea, 212
Vernonia cinerea, 132, 145
    patula, 132
Veronica, 87
Vi, 151, 188
Vigna marina, 36, 43
    sinensis, 148, 228
Vines, ornamental, 160
Vitex negundo, 228
    trifolia, 32, 33, 37, 47

W

Waltheria americana, 128, 140
Waríngin, 158
Watermelon, 228
Wax gourd, 149
Wedelia biflora, 37, 48
Weeds, 118
Werámo, 190

X

Xanthosoma violaceum, 149
Ximenia americana, 31, 42
Xylocarpus granatum, 52, 57
    moluccensis, 31, 44

Y

Yam, 149
Yam bean, 148
Yautia, 149